Wildlife, Landscape Use and Society

A comprehensive analysis of the various terrestrial natural landscapes and habitats within Japan, and the efforts to conserve them and sustain landscape services.

In 2011, Conservation International designated the Japanese islands collectively as one of the world's biodiversity hotspots. They are not only rich in biodiversity, but also densely populated and so human impacts have led to many species being classified as endangered though few have become extinct during recent decades. Sugimura evaluates the effects of landscape changes, government policies and economy on the forest ecosystems and services of Japan. He then contemplates how a rich variety of wildlife species have been able to survive, albeit in limited numbers, despite the rapid expansion of Japanese economic activities in the 20th century. In addition, there appear to be correlations between uniqueness of biodiversity, types of landscape use and the attitudes of local communities towards natural landscapes.

A vital introduction for international environmentalists, geographers and environmental scientists looking to understand Japan's unique ecosystems and their experiences with human activities.

Ken Sugimura is a former professor of Environmental Science at Nagasaki University, Japan (he retired in March 2020).

Routledge Contemporary Japan Series

84 **Masculinity and Body Weight in Japan**
Grappling with Metabolic Syndrome
Genaro Castro-Vázquez

85 **New Frontiers in Japanese Studies**
Edited by Akihiro Ogawa and Philip Seaton

86 **Social Change in Japan, 1989-2019**
Social Status, Social Consciousness, Attitudes and Values
Edited by Carola Hommerich, Naoki Sudo and Toru Kikkawa

87 **Avant-Garde and Nondominant Thought in Postwar Japan**
Image, Matter, and Separation
Kenichi Yoshida

88 **Wildlife, Landscape Use and Society**
Regional Case Studies in Japan
Ken Sugimura

89 **Women and Political Inequality in Japan**
Gender Imbalanced Democracy
Mikiko Eto

90 **Somaesthetics and the Philosophy of Culture**
Projects in Japan
Satoshi Higuchi

For more information about this series, please visit: *www.routledge.com/Routledge-Contemporary-Japan-Series/book-series/SE0002*

Wildlife, Landscape Use and Society

Regional Case Studies in Japan

Ken Sugimura

Routledge
Taylor & Francis Group
LONDON AND NEW YORK

First published 2021
by Routledge
2 Park Square, Milton Park, Abingdon, Oxon OX14 4RN

and by Routledge
52 Vanderbilt Avenue, New York, NY 10017

Routledge is an imprint of the Taylor & Francis Group, an informa business

British Library Cataloguing-in-Publication Data
A catalogue record for this book is available from the British Library

Library of Congress Cataloging-in-Publication Data
Names: Sugimura, Ken, author.
Title: Wildlife, landscape use and society: regional case studies in Japan /Ken Sugimura.
Description: 1 Edition. | New York: Routledge, 2020. | Series: Routledge contemporary Japan series | Includes bibliographical references and index. | Summary: Provided by publisher.
Identifiers: LCCN 2020033675 (print) | LCCN 2020033676 (ebook) | ISBN 9780367333812 (hardcover) | ISBN 9780429319532 (ebook)
Subjects: LCSH: Regional planning–Environmental aspects–Japan. | Forest ecology–Japan. | Urban ecology (Sociology)–Japan.
Classification: LCC HT395.J3 S934 2020 (print) | LCC HT395.J3 (ebook) | DDC 577.30952–dc23
LC record available at https://lccn.loc.gov/2020033675
LC ebook record available at https://lccn.loc.gov/2020033676

ISBN: 978-0-367-33381-2 (hbk)
ISBN: 978-0-367-63496-4 (pbk)
ISBN: 978-0-429-31953-2 (ebk)

Typeset in Galliard
by KnowledgeWorks Global Ltd.

Contents

List of tables

List of figures

Preface

When I was little, my father bought a photography book titled National Park. The book began with the sentence: "The coastline of Japan is as beautiful as the Italians, and its mountains are like the Swiss Alps." It was when the country's high economic growth began, and the Japanese were proud of the beauty of nature rather than their growing economic strength. Since then, human activities have had significant impacts on natural ecosystems and biodiversity, while Japan became one of the world's top economic powers.

On an island of exceptionally unique forest biodiversity, people both inside and outside the island seem to realise its value decades after they began explorations of forest resources that significantly impacted the biodiversity. In another region, home to many of Japan's endemic wildlife species, local people realised the value of biodiversity decades after they began large-scale clear-cutting in natural forests that could negatively impact biodiversity. The national government uniformly applied the forest policy to extend single-species tree plantations across the country. However, common to these regions, unfavourable climate and site conditions prevented them from having extensive areas of commercial plantations. Later conservation activities contributed to protecting the unique biodiversity in these regions. In another region favourable conditions allowed it to have extensive areas of plantations. Over there, there are not many distinctive species, and forest managers are not concerned with conservation. Forests have been used primarily for wood production. Some people like nature very much, while others are relatively indifferent to it. Such differences may have been partially created by the quality of nature that has substantial variation among regions.

Basically local communities are to decide whether to prioritize biodiversity or human activities for every piece of land. On the other hand, contemporary politics and economy are predominantly concentrated in the Tokyo metropolitan area. People tend to have uniform perspectives and ways of thinking, many of which come from Tokyo. For the purpose of promoting local economy based on forestry, the government implemented uniform policies across the country. The idea was simple and easy to grasp but various local factors interacted with each other in reality. Scientists like ideas that they believe widely applicable, such as traditional cultural landscapes that are essential to the richness of Japan's biodiversity. This book goes in a different direction. I found no simple phenomena common throughout different regions but it was not hard to understand what occurred in a given region and differences between regions.

The primary motivation for writing is, except for satoyama, that there is not much literature published in English about human-nature interactions in Japan. I wrote the book entirely by myself, so I owe the responsibility for all the writings and drawings. Nevertheless, this work was not possible without substantial assistance from my friends and colleagues. I sincerely appreciate their kindness, as mentioned in Acknowledgements.

Acknowledgements

I would like to express my thanks to a large number of people who made this book possible and exclusive thanks for the critical reviews and editing of this manuscript to Nana Satake (Chapter 1), Ian Penna (Chapters 2 and 3) and Theodore Howard (Chapters 4 and 5).

In addition to people mentioned in the text, there are many more people to whom I must express my thorough gratitude for their considerable contributions to data collection and processing.

Amami Oshima is an island I visited for my Ph.D. study, followed by various kinds of research activities which extended more than 30 years. I am indebted for field data collection particularly to Yoshitaka Takatuski, Hidemi Kawaguchi, Mikio Takashi, Yumiko Nagai, Yuko Abe, Hisahiro Torikai, Yukari Handa, Fumio Yamada, Asako Miyamoto, Hideki Saito, Shigeho Sato, Shintaro Abe, Takao Shima, biology students from Kagoshima University and many colleagues at the Shinrin sōgō-kenkyūsho (Institute of Forest Sciences). Special thanks must be given to Shosaku Hattori who provided me with the necessary equipment and precious knowledge during the first year of my fieldworks. I would like to express my gratitude to many local government officers who gave crucial information on forestry policies and statistics. Special thanks should be given to those in the forestry section of Uken Village, Amami City and the Oshima-shicho (the branch office of Kagoshima Prefecture) as well as the Amami National Forest Office. I would like to thank the Japan Wildlife Research Centre for conducting the large-scale Amami rabbit studies in the early 1990s. I thank Nobuo Ishii, who was the project leader and also gave me valuable comments on the two chapters.

I had no personal connections in Tadami, but I seldom had difficulty in finding cooperation from the beginning. I was able to conduct the feasibility study with Isao and Fumiko Yamagishi, who allowed me to accompany when they gathered wild plants and mushrooms. Members of Fuzawa Mori-no-ōendan and Asahi Suketto Club joined the project to provide information on their activities in forests. I am grateful to Kokichi Kariya for the provision of local traditional knowledge, to Isamu Nikkuni for the provision of several scientific documents and knowledge on nature conservation activities, to Kishiro Nakamura for coordinating fieldworks in the national forests and to Masumi Nikkuni for providing the population data of the town. I am also indebted to Nobuo Hirano for the questionnaire study in Hinoemata, officers at Tadami Town Office for various kinds of arrangement for the projects and Ryoichi Igari for the provision of Caesium measurement data in Kawauchi. Toshiya Matsuura was a research colleague most of the time, who kindly reviewed the chapter. I thank Mai Sakai, who worked for the data entry at Nagasaki University.

I conducted all the fieldworks in Tsukuba by myself. The analyses were not possible without data entry by Toshiko Kasai and the measurement on the maps by Kimiko Nishibu and the GIS work provided by Toshiya Matsuura.

I am indebted to many people for the works in West Java: staffs at Gunung Halimun Salak National Park (GHSNP), Centre for International Forestry Research (CIFOR) and Japan Environmental Education Forum (JEEF) as well students from Bogor Agricultural University whose names are listed on the referenced paper. I especially thank Popi Astriani and Rosita Go for the assistance at CIFOR and Makoto Yata at JEEF for the work in GHSNP.

These works would not have been possible without the financial support of the East-West Centre (USA), Nihon Seimei Foundation, Sumitomo Foundation, Asahi Foundation, WWF Japan, Kankyo-sho (Ministry of the Environment), Gaimu-sho (Ministry of Foreign Affairs), Nagasaki University and Shinrin sōgō-kenkyūsho.

I would like to express my sincere gratitude to Simon Bates, who visited me in Nagasaki and offered a lot of support and encouragement before the start of this book project. I also thank the editors of Taylor & Francis for their patience to wait for delayed revisions. I finally need to thank my supportive wife, Kumiko, and sons, Ryo, Yu and Takashi, for the generosity of my frequent absences from home during the field surveys.

Introduction: aims and scopes

Backgrounds and themes

Japan is rich in biodiversity relative to its land size and despite its high population density. There are many species endemic to Japan, many on the national Red List. Species' uniqueness and state of endangerment have great regional variety. Biodiversity by definition includes genetic, species, ecosystem and landscape diversity, but I focus here on species diversity. Japan has a great variety of living species from the south to the north owing to the range of climatic and biogeographical conditions, as described in Chapter 1. The main foci of this book are terrestrial vertebrates, in particular mammals and birds, and forest landscapes, which cover about 70% of the land area. Many species have conservation values. As Japan is one of the most densely populated countries and population per arable land area was the highest in the world about half a century ago; most organisms have had to maintain populations under development pressure or in harmony with human activities.

Similar to species diversity, the utilisation of ecosystems by local people has great regional variety across the country.

I divide the country into four regions in terms of characteristics of biodiversity and relationship with human society (Chapter 1). Roughly speaking, three of the four regions, each with distinctive features, represent the uniqueness of Japan's biodiversity and its relation with human society as a whole. The fourth region makes relatively small contributions to the unique biodiversity, so it is placed in the first section of the last chapter to reveal its difference from the other regions. Selecting a case study area in each region, I describe the characteristics and relationship from the following perspectives (Chapters 2–6). First, I evaluate the uniqueness of vertebrate species diversity in terms of the number of species endemic to Japan and the Japanese Red List category to which some of the endemics belong. I adopt the Japanese categories, as explain later. Such an evaluation can help to explain how Japanese biodiversity has been formed and maintained overall, and the differences between the regions.

Second, forest often plays a vital role as it is the largest component of the terrestrial landscape(1) and has changed over the years under the influence of economic background and conservation pressures. The Forestry Agency of Japan started exploiting natural forests in areas remote from human settlements according to rigid national policy determined in the 1960s. Wildlife species that are dependent on old growth must have suffered from losing large areas of their crucial habitat. Nature conservationists fought against such felling across the country before a fundamental change in the policy took place in recent years. International trade in woodchip has also had a significant impact on national forest management policies. Northwestern Kyushu, in the fourth region, is an exception, as exemplified in Chapter 6, where plantations managed for timber production predominate over naturally regenerated forests in most municipalities.

Third, relations between humans and nature vary significantly among the four regions and have changed considerably over time in terms of how people use forests and natural environments, locally perceived benefits and costs of wildlife species, and public concern with biodiversity, among other factors.

Finally, I focus on the distinction between satoyama (rural landscapes) and okuyama (remote mountains). My definitions may generate controversy, as they differ from those in two other books written in English (Takeuchi *et al.*, 2003; Duraiappah *et al.*, 2012). In the traditional view, satoyama is a rural landscape composed mainly of forests, paddy and other fields, settlements, small rivers and ponds. It is a kind of cultural landscape, heavily utilised and extensively distributed throughout the country but for Hokkaido. In contrast, okuyama lies farther from human settlements than satoyama and is sparsely inhabited and mostly covered by forest. It features extensive natural forests with many old trees in the central and northern parts of Japan. Discussion of satoyama is essential for biodiversity conservation, since not only the authors of the above books but also the Ministry of the Environment emphasise that it is the most critical landscape for biodiversity conservation. However, I cannot draw a clear boundary with okuyama. In Chapter 5, I describe the history and current status of satoyama, along with a case study of its biodiversity in comparison with the other case study areas, which I think are more important than satoyama.

Biodiversity is an ambiguous term and biologists dare not use it in case studies. I argue here that it means species diversity across the country, comparing one region with another and demonstrating ideas through concrete examples rather than discussing them as concepts.

Landscape changes during the 20th century

In the past, when the main fuel sources were charcoal and firewood instead of fossil fuels, people visited nearby forests daily and walked far rarely. Frequently used forests abound in the mountains around the city of Nagasaki, where I lived for 5 years and enjoyed walking in historically coppiced satoyama forests (Chapter 6). In satoyama, local people used to cut young trees to encourage the growth of several new stems from the edge of the stump. Coppice forests, featuring many such trees, are easy to distinguish from naturally regenerating secondary forests. The people used to manage large areas of coppice forests to obtain manure from leaves as well as charcoal and firewood (Saito and Shibata, 2012; Chapter 5).

In the 1960s, when economic growth was the fastest in Japan, manufactured fertilisers replaced manure and coal replaced wood as fuel. Now, coppice forests have been replaced by even-aged single-species plantations, unmanaged naturally regenerated forests, farmland, resorts or, most recently, solar panels.

The Japanese archipelago has many mountains with steep slopes, which historically hindered intensive forest use in remote areas. There are such mountainous forests on the island of Amami Oshima and in the western Fukushima Prefecture (Aizu), which are far from residential areas and have had large-diameter trees in the past (Chapters 2 and 4). However, the introduction of vehicular access and mechanised forestry facilitated rapid forest development in okuyama, where large trees were rapidly lost and plantations replaced natural forests. Chapters 2–4 describe how the forests in okuyama underwent considerable changes, unlike in satoyama, where local people did not use the forests daily and many large-diameter trees grew. From the 1960s to the 1990s, forest managers built logging roads to allow them to clear-cut the old broadleaf forests for pulpwood and to convert some of them to conifer plantations. Such operations caused significant changes not only in plant species and forest age composition, but also in habitat conditions for many animal species. As this extensive loss of old natural forests drew to an end, Japanese society entered into an era of a bubble economy. This had significant effects on forests for about 5 years around 1990, with the clearance of large areas for recreational facilities, such as ski slopes and golf courses. Government protected only small areas from these impacts (Chapter 1).

National and local governments were ignorant of the need for biodiversity conservation until 1992, when Japan attended the Rio de Janeiro Earth Summit. This was a turning point, when the Cabinet of Japan declared that it would play a leadership role in problem-solving, albeit without understanding the global environmental problems. Most of the laws on biodiversity conservation were enacted after this year, facilitating a fundamental change in forest laws and the enforcement of Environmental Impact Assessment Law.

Literature review

There are many studies and scientific reports from limited disciplines that describe and argue for biodiversity conservation and ecosystem services within specific geographic areas. But there are few overviews even of biogeography but for those published more than 40 years ago (e.g., Tokuda, 1969). Chapter 1 summarises biogeography, focusing on that of terrestrial vertebrates as a representative group, before presenting case studies for comparison between regions in the following chapters. One of the few general overviews of biodiversity and ecosystem services (Kankyō-shō, 2012) has some fundamental differences from the arguments in this book. First, it considers Japan as a single entity and evaluates it using many national statistics. In contrast, through case studies, I approach issues of geography, ecology, conservation and local society from a broad perspective. Second, the 2012 report involves all ecosystem types, while this book focuses on forests, except that Chapter 5 deals with satochi landscapes[2]. Third, the 2012 report separates biodiversity and ecosystem services, whereas I describe in a more integrated manner how animals and people use landscapes in a region, and discuss differences between areas in the relationship between people and landscapes. Fourth, the 2012 report divides ecosystem services into four categories—basic, provisioning, regulating and cultural—to describe them separately on the basis of many country-level statistics. On the other hand, I focus on ecosystem services that depend mainly on forest biodiversity, such as mushroom and wild vegetable harvesting, stream fishing and nature tourism. These services often fall into two or more categories simultaneously, so it is not practical to treat the categories separately. For example, people distinguish edible and poisonous mushrooms through traditional knowledge, which is a kind of recreational food culture (provisioning and cultural services at the same time). As Wallace (2007) suggested, a case study on ecosystem services should not separate the four categories but integrate them in practice.

Little progress has been achieved in studies of biodiversity-based forest ecosystem services in Japan, despite large numbers of both ecologists and forest managers. Ecologists have published many articles regarding endangered species on Amami Oshima, with a focus on species ecology and biological conservation (Chapter 3). Forest managers, in contrast, are concerned with wood production in even-aged, single-species plantations, so few studies deal with interactions between forest management and biodiversity conservation from a socio-ecological point of view. Therefore, these two groups do not have the opportunity to work together. Nevertheless, Sugimura (1987) and some others did integrated studies of forest management, species conservation and ecosystem services, as introduced in Chapter 2. Chapter 3 discusses a national park on Amami Oshima, which was the first in Japan established primarily for biodiversity conservation, along with another in Okinawa. Ito (1996) reviewed the influence of forestry on the formation of national park policy in Japan, but he wrote it before biodiversity played a vital role in the establishment of national parks after the 1992 Earth Summit. Fortunately, I had an opportunity to follow such a process on Amami Oshima. Sugimura (2016) made some suggestions on sustainable forest use and conservation based on results obtained from recent work on tourism and forest management, including regulations set in the new national park, as illustrated in Chapter 3.

No study has reviewed forest cultural services in Japan except for that of Saito (2006), who developed a database and identified from a statistical analysis more frequent harvesting of wild vegetables and mushrooms in the northern part of the country. I used the results of this analysis to draw the boundary between regions where local people often harvest non-timber forest products (NTFPs) and where recreational activities are most frequent. Shackleton *et al.* (2011) did an extensive review of NTFP in the global context. They pointed out that Sweden is one of a small number of exceptional cases among developed countries where enjoying NTFP harvest is common, but Japanese cases were unknown to them. Some researchers have paid particular attention to forest cultural services, such as recreational stream fishing, hunting and tourism, based on surveys conducted in and near Fukushima Prefecture. For example, Kitamura *et al.* (2004) and Tamaki *et al.* (2016) studied anglers' behaviour in the south of Fukushima.

The consequences of the accident are significant, as reported in many publications (Chapter 5). Many reports are concerned with the accumulation and circulation of caesium in forest ecosystems, measuring radioactivity in soil, water, plants and animals. On the other hand, there are few reports of the effects on forest ecosystem services except for wood production. I was involved in a research project designed to quantify forest ecosystem services from 2008 to 2011, right before the earthquake. Kawarasaki and Sugimura (2012) and Sugimura (2012) carried out nationwide statistical analyses on forest services, revealing some regional characteristics of western Fukushima Prefecture. After the accident, I restarted a similar study to find out the effects on forest services since 2014. I compared the data before and after the accident to assess the impacts on various kinds of services in Fukushima.

Many articles on satochi and satoyama landscapes deal with ecosystem services and biodiversity (e.g., Takeuchi *et al.*, 2003; Duraiappah *et al.*, 2012). The landscapes are located between urban areas and remote mountains (okuyama) and are maintained by human activities (Akimichi and Morimoto, 2012). As they are composed of a variety of ecosystem types, they have high levels of species diversity (e.g., Katoh *et al.*, 2009). At the same time, they can potentially provide a variety of ecosystem services, since population density is high relative to okuyama. Chapter 5 introduces studies of biodiversity and ecosystem services in such landscapes. One of the problems with satoyama is that its definition and geographical boundaries are not clear (Saito and Shibata, 2012). For example, Saito and Shibata (Ibid) suggested that coppice forests and reservoirs were substantial components of satoyama, where harvesting of wild vegetables and mushrooms, prescribed burning of grassland and collecting fallen leaves for manure production are commonly observed. Yet such landscape management and utilisation do not exist anymore (Akimichi and Morimoto, 2012). Takeuchi (2003) suggested that conservation strategies should include coppiced woodland. These satoyama advocates maintain that the disappearance of traditional management is the major reason why there are now so many endangered species in satoyama. On the other hand, Tsunekawa (2003) defined satoyama and satochi together in terms of terrain (slope angle), human population density and percentage of woodland cover, and estimated that satoyama landscapes constituted approximately 20% of the national land area. But it is no longer possible for satoyama landscapes, which are centred on coppiced forests, to cover such large areas. Thus, reality and logic do not align with each other very well in these arguments. Another problem is that the satoyama advocates do not take the Red List categories into account, yet they suggest that there are more endangered species than in any other type of landscape. I introduce some local examples in Chapter 5 and discuss some of the logical contradictions and the actual value of satoyama. Contrary to their arguments, Chapter 2 shows that old natural forests provide suitable habitats for many endangered endemic species, and Chapter 4 shows that okuyama hosts many species of mammals endemic to Japan.

Objectives

Conservation International (2011) has designated the Japanese Islands (the whole country) as a biodiversity hotspot. This designation suggests two interconnected but different aspects: that Japan is rich in biodiversity and that human impacts have been so high that there are many endangered species. As described in the following chapters, Japan currently has many more surviving endangered species than extinct species. So how have they survived, while no effective measures for preservation have been practised until recently, as implied above? The reasons and the situation vary significantly by region, as I show by dividing Japan into four regions. I speculate on this question in terms of landscape changes, species composition, government policy and economic background.

I have two types of species with conservation values in mind. One is the species that have survived in satoyama and must therefore have adapted to anthropogenic habitats, which feature many coppiced trees and few old trees. Game species and harvested plants may well have been at occasional risk of local extinction under excessive hunting or harvesting pressure. Later, many of them had to live through agricultural pollution from the use of manufactured fertilisers and pesticides and competition with invasive species. The current and the past satoyama are very different, and they have had to survive through such significant changes in habitat conditions. Chapter 5 examines the characteristics of species composition in satochi in terms of Red List categories, endemism and indigenousness, based on a literature review and field observations in Tsukuba City and the Mount Tsukuba area. The other type of species typically lives in okuyama under much less disturbed conditions than in satoyama. One example is birds and mammals that use cavities formed in large broadleaf trees for nesting. These cavity nesters need many large trees to sustain their populations, as exemplified in studies on Amami Oshima. Another example is those living at high elevations away from human settlements without experiencing high levels of human disturbance and harvesting intensity.

One thing I would like to emphasise is the importance of adopting a geographical perspective in any nationwide assessment. Next chapter explains the perspective based on four regional divisions, as briefly described below:

(A) Northern Honshu, where deciduous broadleaf trees are dominant
(B) Central Honshu, where population density is high
(C) Southern Shikoku and Kyushu, where single-species conifer plantation is the most extensive land cover type
(D) Remote islands at the south-western end of the country, where many rare species live in evergreen broadleaf forests.

Ueda and Onodera (2004) and Kuramoto and Sonoda (2012) concluded that there were generally more endangered species in satoyama than in okuyama and climax forests. There is far more literature that emphasises the importance of biodiversity conservation in satoyama than in okuyama. Is it certain that okuyama has fewer endangered species than satoyama? I examine habitat preferences of species on the Red List in the case study areas (Chapters 2, 4 and 5) and attempt to identify which type of habitat they prefer to live in. On the other hand, forest biodiversity provides various kinds of value through landscape use, such as recreation, harvesting mushrooms and wild plants, angling, hunting, and opportunities for environmental education, ecotourism and nature observation. There may be regional differences in such uses in terms of the frequency and preferred types of service, which may be associated with the regional differences in biodiversity, which I hope will be evident in this book.

Finally, the Ministry of the Environment is responsible for national policies on biodiversity conservation and has many regional offices for national park management. Chapter 3 describes the substantial role that it plays. Chapters 1 and 6 discuss the issues facing it. Comparison with other countries makes the points more evident. Chapters 2–5 describe how Japan's biodiversity is preserved at the local level or otherwise ignored. The role of local conservation movements has been significant at times and dependent on the strength of incentives that are also different among regions.

This book examines the validity of the following arguments. (1) Three contrastingly different regions, complementing each other, form the richness of Japanese biodiversity and its values, while one region does not make a sufficient contribution. (2) Economy and local awareness, rather than national conservation measures, have had greater impacts on species diversity and endangered species status. (3) Activities to maintain or restore traditional satoyama landscapes are not very effective for biodiversity conservation at the national level. (4) Considerable changes in biodiversity conservation and biodiversity-based services since the 1960s may give us a better future than the past.

Supplementary notes

1) The definition of landscape follows Forman and Godron (1986). It describes it as a heterogeneous land area composed of a cluster of interacting ecosystems, which is repeated in similar form throughout.
2) Satochi landscape is dominated by farmlands, whereas woodland is dominant in satoyama.

1 Terrestrial biodiversity and landscape use across the country

A brief overview

1.1 Geographical landscapes of Japan

The land mass of Japan is about 378 thousand km^2, consisting of four main islands: Hokkaido located in the north, Honshu in the centre, Shikoku and Kyushu in the southwest (Figure 1.1). Off the south shores of Kyushu expands a long chain of islands, the Nansei Archipelago, extending to the west of Taiwan. The archipelago of islands extends in a northeast-southwest direction, with approximately equal length of Honshu. The country is subdivided into 47 prefectures. Japanese people often feel a sense of belonging in a specific prefecture or region, occasionally an island, as a regional unit. The population was the 11th largest in the world in 2019, in contrast to total land mass, which ranks 62nd (United Nations, 2019), indicating high population density.

Honshu is administratively divided into five regions, namely Tohoku (northernmost part), Kanto (east-central, south of Tohoku), Chubu (central, west of Kanto), Kinki (west-central) and Chugoku (westernmost Honshu). Mountain ranges with high altitudes and steep slopes in a row demarcate the boundaries clearly between Tohoku, Kanto and Chubu regions. Chubu has high mountain ranges with some 3000 m class mountains. Its southern part of the border with Kinki is not very clear, located in the second-largest plain in the country. Fukushima Prefecture, which unfortunately became world-famous after the nuclear power plant accident, is located towards the southernmost border in the Tohoku region and just north of Kanto. The national government occasionally places regional administrative offices according to these geographical divisions to implement national policies into practice. In the case of environmental policies, an example is the Forestry Agency that retains a regional office in Kyushu to manage the national forests and subsidises many projects operated by local governments as well as those in collaboration with private woodlands across the regions of Kyushu and Okinawa. Amami Oshima, which is discussed extensively in the following two chapters, is one of the largest islands of the Nansei Archipelago, belonging to the Kagoshima Prefecture, which is located southernmost in Kyushu, though this region resembles Okinawa geographically, historically and culturally. The Ministry of the Environment, in the Okinawa regional office, declared the importance of the Nansei Archipelago for crucial biodiversity conservation in addition to the main Kyushu region.

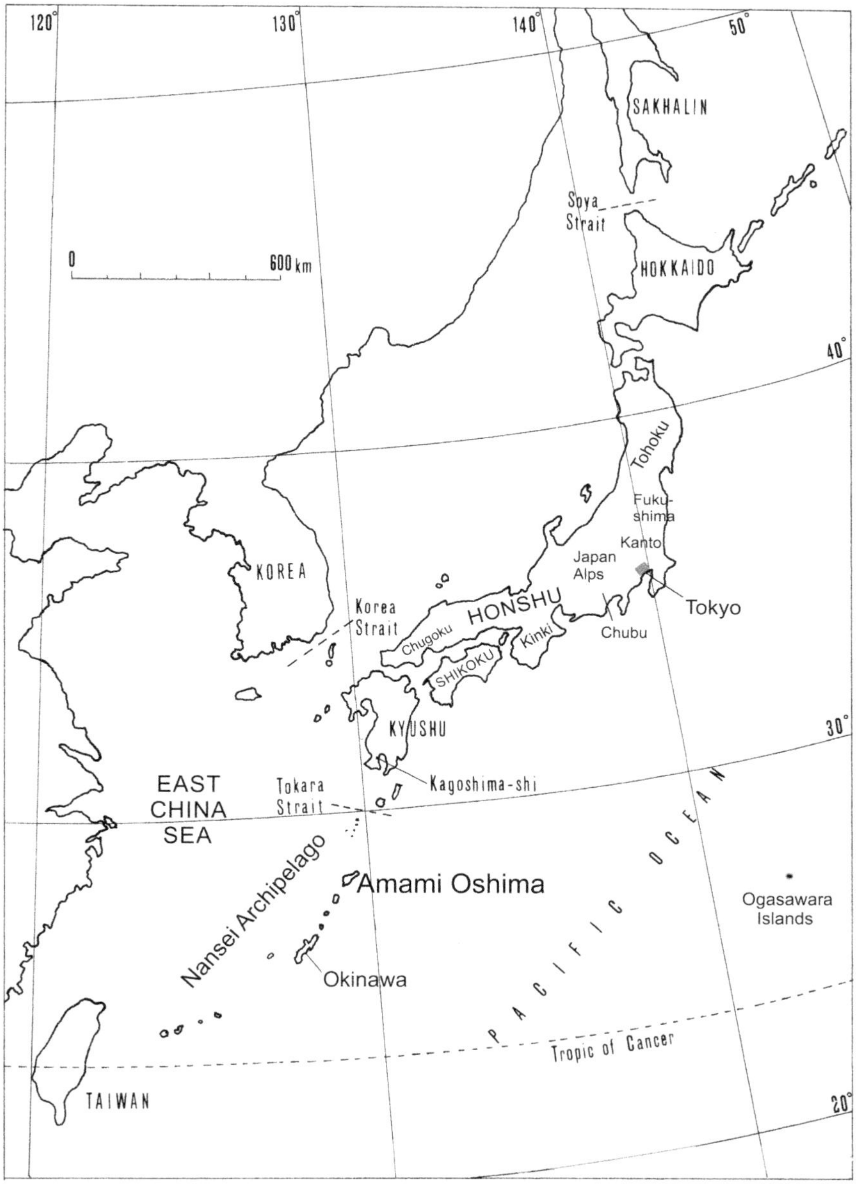

Figure 1.1 Geography of Japan: Names of key regions and locations described in the following chapters.

Population size, density and distribution have large geographic variations. For example, Hokkaido has the lowest density of 64 people/km^2, Tohoku ranks the second lowest of 134 and Kanto has the highest density of 1326 (Sōmushō Tōkeikyoku, 2017). The Kanto Plain, the largest plain in the country, forms a metropolitan area with three neighbouring prefectures, including Tokyo, where the population is about 30 million people, about a quarter of the population of Japan. The second-largest metropolitan area, so-called Keihanshin, is in the western part of Kinki region, where half of the population of the Tokyo metro area lives in cities that have relatively low mountains nearby. Mountainous areas are popular sites for recreation, where people enjoy a good view from the top. The most remarkable difference between the two areas is that a few mountain peaks are far more crowded with visitors in the Tokyo area than in Keihanshin. The use is more dispersed in the latter because there are more mountains nearby. The Pacific Belt Zone has a high concentration of urban populations, extending from Tokyo to the northern part of Kyushu. It covers coastal areas of Pacific Ocean in Kanto and Chubu, the central Kinki and an inland sea (Setonaikai) between Chugoku and Shikoku regions. Excluding Hokkaido and Tohoku regions, the population density of a prefecture is, generally speaking, inversely correlated with the percentage of forested area. In the Kanto region including Tokyo, where population density is the highest at 1326/km^2, forests cover only 43% of the land mass. In contrast, in the three prefectures west of Tokyo, the forests cover is vast at 80% of the land mass with a small population density of 173/km^2.

Natural vegetation and ecosystems are immensely diverse from Hokkaido to the Nansei Archipelago due to the long length of Japan, from the most northern point of 45 degrees to the most southern point of 20 degrees North latitudes. The natural vegetation of Hokkaido represents either cool-temperate or subarctic climates in low-altitude plains, where deciduous broadleaf trees and conifer trees dominate. Hokkaido is cold in winter, and the average temperature in January in Sapporo (43 degrees N latitude), the largest city, is minus 3.6 degree Celsius. The wildlife fauna of Hokkaido is more similar to that of the Eurasian continent to the west than to the south (Honshu) as indicated below. Additionally, this region underwent a radical change in the use of natural landscapes, when the Japanese occupation drove out the indigenous people, the Ainu, during the last half of the 19th century. Consequently, from the traditional land use of natural resources by the Ainu, in the present day, much of the land use is now largely agricultural, and hence Hokkaido has lost much of its traditional use of landscapes, although its reminiscence can still be seen in the Tohoku region. Typically, the climate of the Tohoku region lends itself to deciduous broadleaf forests. These characteristic broadleaf forests vegetation zone gradually moves towards areas of high altitude from southern parts of Tohoku to southern Kyushu, where the zone near the mountain apex, such as that in Yakushima Island, can be above 1500 m high and equivalent to the Tohoku region (Kankyō-shō, 2010). Zones of lower elevations of the southern regions are often dominated by evergreen oaks (*Quercus* spp.), although formally, the natural vegetation would have been broadleaf evergreen forests. This is due to extensive replanting of commercial timber trees, such as sugi cedar *Cryptomeria japonica* and hinoki cypress *Chamaecyparis obtusa*, wherever soil conditions were deemed favourable in this zone. In the Chubu region, the apparent transition from the evergreen broadleaf forest to deciduous broadleaf forest occurs at altitudes of 500–600 m (Otsuka and Ozeki, 2015). In contrast, the Nansei Archipelago has canopy tree species not found in the mainlands, such as *Schima wallichii* and *Distylium racemosum*, suited to the subtropical to tropical climates. Epiphytes are common, and hence the forest interior looks similar to montane forests in Java (Chapter 6). Mangrove forests grow in coastal swamps, which are not found in mainland Kyushu.

1.2 Species richness

Japan is rich in biodiversity and has a broad range of ecological niches relative to its land mass. Four major factors contribute to species richness:

(1) Wide range of latitude from Hokkaido to Okinawa
(2) The elevation difference from sea level to mountain summits 2000–3000 m high
(3) The connections of Japanese islands to the Asian continent during the last glacial periods by land bridges
(4) Combinations of natural and anthropogenic disturbances

Japan extends from Cape Soya in Hokkaido, at 45 degrees 31 minutes N Latitude, to Yonaguni Island as the southwestern tip of the country, which is located east of Taiwan at 24 degrees and 27 minutes N Latitude. Two opposing ocean currents, a cold stream (Oyashio) and a warm stream (Kuroshio) flowing from the north and south, respectively, amplify the climatic differences. The Oyashio Cold Current flows northeast of Hokkaido to the south along the Pacific coast, which is strong in winter and cools Hokkaido down further contributing to the relatively cooler temperate conditions in the north. The Kuroshio Warm Current flows northward off the Pacific coast, and its strong summer current makes the Japanese summers significantly hotter and tropical. Thus, the climate differences between the north and the south are more distinctive than that of countries in Western Europe or the west coast of North America with similar land mass. The average temperature ranges from minus 3.7 degree Celsius in January and 17.3 in August in Nemuro, a city in the easternmost part of Hokkaido, while it is 18.4 degree Celsius in January and 28.5 degree Celsius in August on Yonaguni Island as the latest three-decade average (Kishō-chō, 2019).

The vast differences in climatic conditions are reflected in the differences observed in the dominant canopy tree species between Hokkaido and the Nansei Archipelago. Okitsu (2003) reported that a boreal forest had conifers such as *Picea jezoensis*, *Picea glehnii* and *Abies sachalinensis* as well as deciduous *Larix gmelinii* as dominant species in undisturbed forests in central Hokkaido. Hiura *et al.* (1995) found that *Abies sachalinensis*, *Picea jezoensis*, *Quercus mongolica*,

Sorbus commixta, *Acer mono* and *Kalopanax pictus* were dominant in the primeval boreal mixed forest in northern Hokkaido. In direct contrast, *Castanopsis sieboldii*, *Schima wallichii* and *Distylium racemosum* were dominant in the evergreen broadleaf forests of Okinawa Island (Xu *et al.*, 2001). Slightly north of the islands to the mainland Kyushu region, other canopy species, such as *Quercus miyagii*, *Myrica rubra*, *Cinnamomum camphora*, *Schefflera heptaphylla*, *Ilex liukiuensis* and *Machilus thunbergii*, are reported dominant (Takashima, 2019). Mangrove forests in this region have three species: *Bruguiera gymnorrhiza*, *Kandelia obovata* and *Rhizophora mucronata* (Kawamitsu *et al.*, 1997).

Central Honshu has the highest mountains in Japan, the so-called Japan Alps, which has three separate mountain ranges with about 150 peaks over 2500 m high. Fortunately, anthropogenic activities have not directly altered the unique fauna and flora at the highest altitudes and the region has been retained as the most popular recreational nature tourism destination from the Tokyo metropolitan area. No studies have summarised regional characteristics of biodiversity and nature tourism within the relevant literature. The flora of the alpine zones is characterised by the dominance of Siberian dwarf pine, *Pinus pumila*, which grows only up to about knee height (Photo 1.1). Severe environmental conditions for organisms have maintained unique species, which have fragmented distributions in high-altitude mountain ranges. The most well-known and popular species living within the dwarf pines is a nearly flightless bird, Japanese rock ptarmigan *Lagopus muta japonica*. This subspecies occupies the southernmost latitudes of the *L. muta* spp. distribution range, compared to subspecies in the Eurasian Continent (IUCN, 2019). Another nationally well-known and unique

Photo 1.1 High mountains covered by dwarf pines.

species within these alpine zones is *Callianthemum hondoense*, a plant of the buttercup family. It is found only near the top of Mt. Kitadake, above 2750 m, in the South Japan Alps (Natori, 2008). The ptarmigan and *Callianthemum hondoense* are ranked second and third highest in the national Red List, and hence endangered and vulnerable, respectively (Kankyō-shō, 2018).

There are numerous mountains with a peak height of greater than 1700 m across Japan, from Hokkaido to Yakushima. The rich biodiversity in central Honshu has been attributed to the presence of many endemic species and subspecies of the flora and fauna in addition to the altitudinal variation of ecosystems. The mountain ranges on Honshu Island, is the only region in Japan that demonstrates the aforementioned four distinctive vertical vegetation zones. Natural forests in the lowest zone, below 1000 m, are dominated by evergreen broadleaf trees, many of which belong to the oak family (Murata, 2005). Deciduous trees represent the area between 1000 and 2000 m, such as Japanese beech *Fagus crenata* and Japanese oak *Quercus mongolica*. Just prior to the Alpine zone, evergreen hemlock and fir trees are dominant, such as *Tsuga diversifolia* and *Abies veitchii* (Kurita, 1983). Many scientists indicate that numerous relic species had much wider distribution ranges during the last glacial period. As the populations declined, through habitat loss and fragmentation, many have become endangered or threatened, which has permanently changed the ecosystems of the mountains. One of the most significant threats of modern times is global climate changes. Warming of the Alpine regions threatens the survival of biodiversity that inhabit the Alpine peaks of high mountains, such as *Callianthemum hondoense*, a small herbaceous plant (Natori, 2008).

The sea level dropped more than 100 m during the last glacial Ulm period, which created three land bridges between the Japanese islands and the Asian Continent (Oshima, 1991). These land bridges considerably influenced the formation of the current fauna and flora of Japan. The three land bridges connected (a) Hokkaido and Far East Russia through Sakhalin Island, (b) western Japan and eastern China through Korean Peninsular and (c) southwestern end of the Nansei Archipelago and southern China through Taiwan. The magnitude of these effects of these land bridges can be observed in the distribution of some large terrestrial animals, such as bears. The separation of the Hokkaido and Honshu islands by the Tsugaru Strait, and henceforth hindering migration, distinguishes the distribution areas of the brown bear, *Ursus arctos,* and Asian black bear, *Ursus thibetanus*, between Hokkaido and Honshu-Shikoku regions, respectively. Two subspecies of Bengal cat are separated only by a land bridge, one *Prionailurus bengalensis iriomotensis* on Iriomote Island located to the east of Taiwan and the other *P. b. euptilura* on Tsushima Island on the south of the Korean Peninsula. Japanese monkey *Macaca fuscata* and Japanese serow *Capricornis crispus* live on the northern tip of Honshu, Shimokita Peninsula, distributed throughout the mountains of Honshu, Shikoku and Kyushu but are absent in Hokkaido and the Nansei Archipelago. As described in the following chapter, deep sea straits have isolated the central part of the Nansei Archipelago from the continent for at least several hundred thousand years.

Natural and human disturbances have also contributed to maintaining high levels of species diversity at a landscape level. Natural disturbances often remove or change ground vegetation and attributes of the surface soil, enabling development of an intricate pattern of grasslands and forests. Such disruptions may be caused by severe weather events such as floods, landslides, fires, strong winds, avalanches, earthquakes and volcano eruptions. Among these, the more severe events, such as typhoons, which can bring extraordinary volumes of heavy rain, can consequently cause additional sequential afflictions such as landslides. Partial replacement of forested areas with grassland can create new habitats and ecosystems. Additional herbivorous animals feeding on new forest edge plants within the grasslands and swamps enable an extension of hunting zones for raptors nesting in the forests. Raptors nest in large trees and hunt their prey in grasslands and swamps. Stretches of land between the foot of the mountains and flat arable land, known as satoyama, is a relevant example of human disturbances (Chapter 5), which has created and maintained a mixture types of anthropogenic ecosystems around rural settlements to increase diversity of habitats and niches.

1.3 Endemic and endangered fauna and flora

Isolation of the Japanese islands from the Asian continent promoted speciation after the elevation of sea level has produced many endemic species and subspecies within the national fauna (Table 1.1). The total number, indicated as "All species" in the table, includes non-native invasive animals, including the black rat *Rattus rattus* and introduced game bird, Chinese bamboo pheasant *Bambusicola thoracicus*. The rate of endemism, discussed hereafter, would be higher if non-native species were excluded. Amphibians have the highest endemism. Dominant groups of endemics are 17 of 18 (94%) salamander species (family Hynobiidae and Cryptobranchidae), all the three species of newts (Salamandridae), 26 of 37 (70%) species of frogs and toads (Hylidae, Ranidae,

Table 1.1 Number of terrestrial vertebrate species and red-listed species endemic to Japan

Classification	*All species*	*Extinct species*	*Endemic species*[a]	*Endemic species in the Red List*
Mammals	116	2	48	22
Birds	633	5	16	9
Reptiles	80	0	46	23
Amphibians	58	0	46	42

Source: Kokuritsu Shizen Kagaku Hakubutsukan (2018), Motokawa *et al.* (2006), Nihon Yachō-no-kai (2012), Nihon Hachū-ryōseirui Gakkai (2019), Nihon Bunruigaku Rengō (2003), Kankyō-shō (2018).

Note:
a The number of species in this column excludes extinct species.

Rhacophoridae, Microhylidae and Bufonidae). More than half of the reptiles are endemics. Endemism is particularly high among the two families of lizard (Scincidae and Lacertidae), 15 of 21 (71%) species across the two families. Endemism is high among the common snakes (Colubridae spp.), 13 of 19 species (68%), and viper snakes (Viperidae spp.), 6 of 7 species (86%). Naturally, birds, the most mobile group (class), have the lowest rate of endemism. No particular taxonomic group of birds has more endemics than the others, while their geographical distributions are uneven. High endemism among the mammals is limited to the three orders of mammals with small body size, 14 of 20 (70%) insectivorous species (Soricomorpha spp.), 16 of 37 (43%) bat species (Chiroptera spp.) and 12 of 26 (46%) rodent species (Rodentia spp.). The Nansei Archipelago itself is particularly unique as a subdivided region of Japan. It has six endemic species, and the Ogasawara Island group has five. The central part of the former archipelago has been isolated from the continent for an extended period, and the latter is an oceanic island group. Stattersfield *et al.* (1998) also noted the high avian endemism of these island groups based on a quantitative evaluation index. Nansei Archipelago has many endemic species of mammals, reptiles and amphibians as well as birds, as described in Chapter 2.

High-altitude mountain ranges are another typical geographic division that generates high levels of endemism. Ebihara (2011) compared the areas of high endemism of vascular plants at the national level and ranked each area according to the total value obtained by summing the score for each endemic plant living in the area. As a result, similar to the above conclusion, islands occupy the top four positions and mountainous regions occupy the fifth through tenth positions.

Many of the endemic species are endangered. Table 1.1 indicates that all the endemic amphibians but four are categorised as endangered or threatened in the "Red List of Threatened Species of Japan". Table 1.2 shows a list of the endemic species of higher categories in the Red List, i.e., critically endangered (CR) and endangered (EN). The Red List status is according to Kankyō-shō (2018). Information on the distribution was obtained from Abe (2008) and Kankyō-shō (2018). There are some notable features of geographic and taxonomic perspectives that have influenced these outcomes. First, over half (18 of the 39) the species in the table are only found in the central Nansei Archipelago, indicating that the absence of the land bridges has played a crucial role. There are exceptions, namely the salamanders, a mole and a frog. All the other species live either in the Nansei Archipelago or the Ogasawara Islands. This emphasises the significance of the regions and should have the highest priority for conservation. Chapter 2 discusses in detail the habitat changes, population trends and conservation efforts for some species living on one of the Islands, Amami Oshima. Second, in contrast to the above areas, the main islands of Japan are deficient of endemic fauna. Some fauna species in Hokkaido have close relatives on the continent; nevertheless, there are not many endemic species. This can be attributed to the extensive areas of conifer tree plantations in Kyushu, southern Shikoku and southern Kinki, which limits species diversity. Chapter 6 reports some results of bird survey in northern Kyushu to confirm this intuitive reasoning.

Table 1.2 Endemic species in the higher categories in the Red List

Common name[a]	*Scientific name*	*Category*	*Approximate locations of distribution*
Shrew	*Crocidura orii*	EN	Central Nansei[b]
Mole	*Mogera etigo*	EN	North-central Honshu
	Mogera uchidai	CR	Island east of Taiwan
Bat	*Pteropus pselaphon*	EN	Ogasawara
	Rhinolophus pumilus	EN	Central and western Nansei
	Nyctalus furvus	EN	Central and northeastern Honshu
	Myotis yanbarensis	CR	Central Nansei
	Miniopterus fuscus	CR	Central and western Nansei
	Murina ryukyuana	CR	Central Nansei
Mouse	*Tokudaia osimensis*	EN	Central Nansei
	Tokudaia muenninki	CR	Central Nansei
	Tokudaia tokunosimensis	EN	Central Nansei
	Diplothrix legata	EN	Central Nansei
Rabbit	*Pentalagus furnessi*	EN	Central Nansei
Rail	*Gallirallus okinawae*	CR	Central Nansei
Woodpecker	*Sapheopipo noguchii*	CR	Central Nansei
White-eye	*Apalopteron familiare*	EN	Ogasawara
Old World flycatcher	*Turdus celaenops*	EN	Northern Nansei, Islands north of Ogasawara,
Petrel	*Puffinus bryani*	CR	Ogasawara
Gecko	*Goniurosaurus splendens*	EN	Central Nansei
Lizard	*Plestiodon takarai*	EN	Islands east of Taiwan
Grass lizard	*Takydromus toyamai*	CR	Southwestern Nansei
Snake	*Opisthotropis kikuzatoi*	CR	Central Nansei
	Hebius concelarus	EN	Southwestern Nansei
	Calamaria pfefferi	EN	Southwestern Nansei
Salamander	*Hynobius katoi*	EN	Central Honshu
	Hynobius abei	CR	Central western Honshu
	Hynobius amakusaensis	CR	An island in Kyushu
	Hynobius osumiensis	EN	Southern Kyushu
	Hynobius shinichisatoi	EN	Central Kyushu
	Hynobius hidamontanus	EN	Central Honshu
	Hynobius takedai	EN	North-central Honshu
	Onychodactylus tsukubaensis	CR	Mountain in eastern Honshu
True frog	*Glandirana susurra*	EN	Island in North-central Honshu
	Odorrana splendida	EN	Central Nansei
	Odorrana ishikawae	EN	Central Nansei
	Odorrana utsunomiyaorum	EN	Central Nansei
	Babina subaspera	EN	Central Nansei
	Babina holsti	EN	Central Nansei
Rice frog	*Limnonectes namiyei*	EN	Central Nansei

Abbreviations: CR: critically endangered; EN: endangered.

Notes:

a Common family name or otherwise a group larger than the species.

b Central Nansei indicates that the distribution area had not been connected with the continent during the last glacial period.

Third, as mentioned above, the salamander is only a taxonomic group that has several endemic and higher-ranked endangered species living in Honshu. This is often credited to the abilities of the salamanders to adapt and acclimatise to the many satoyama areas. Hence, many scientists suggest that satoyama should be claimed as the essential habitat type for biodiversity conservation (Chapter 5). The species listed in the table is unsupportive of this claim, as examined later, due to the account of species in the lower categories of the Red List. Consideration should also be taken to more than one thousand vascular plant species high, i.e., CR and EN, in the national Red List ranks, which are too numerous to analyse in this book. Nevertheless, the assessment of vascular plants by Ebihara (2011) is consistent with the observations discussed in this book.

Table 1.1 indicates that seven species have already been extinct, including a species of kingfisher for which it is uncertain if the species had existed or not. The two mammalian species are Japanese wolf (*Canis lupus hodophilax* and *C. l. hattai*) and Japanese river otter (*Lutra lutra nippon*). The four bird species include two pigeons (*Columba versicolor* and *Columba jouy*), a flycatcher (*Cichlopasser terrestris*) and a finch (*Chaunoproctus ferreorostris*). Three of the bird species lived in the small islands of Ogasawara and the other one in the Nansei Archipelago. Another species that has virtually gone extinct is the Japanese crested ibis *Nipponia nippon*. This is a controversial species, currently its extinction status is fiercely debated in the government. There was an introduction of individual birds of the same species from China in an attempt to restore the populations, which has appeared to be successful. Ibis was widely distributed across Japan, before hunting and pesticides reduced the bird populations to a small number of individuals. It survived in the Noto Peninsula and Sado Island (both located in the north end of the Chubu region) before its extinction on Sado Island in 1981. The Japanese wolf, otter and ibis all had lived in the main Honshu Island. They have all become extinct almost exclusively due to anthropologic causes and influences, and the same causes of removal of such keystone species, also explain the contribution of population decline in many other fauna species. Overhunting became a particularly significant cause when in the late 19th century, the government opened the country to foreign trade and local people were able to obtain guns easily to kill wildlife efficiently. Nakazawa (2010) suggested the major cause of wolf extinction from northeastern Honshu was "nuisance killing" to protect livestock; in addition to farmers, the police and local people also participated in this hunting due to the introduction of a bounty system. In 1905, an American zoologist, Malcom Play fair Anderson, bought a specimen of a wolf in central-western Honshu, in the Nara Prefecture, the last of its kind, which currently remains the latest information on this species (Kankyō-shō, 2018). The extinction process was similar to the Ezo wolf in Hokkaido (Walker, 2009). The main cause of the ibis extinction was also hunting, since farmers disliked the birds and considered them agricultural pests. They killed the birds not only for feathers and food but also because they caused inconvenience (Yanai, 2006). The birds moved their habitat from satoyama areas to okuyama areas, until forests in okuyama were decimated after the Pacific War (Ibid).

In the present day, hunting is no longer a factor that may affect the survival of any species in Japan. The national Wildlife Management Act used to aim to regulate hunting, but currently the main task, according to this law, is to reduce the population size of pest species, such as wild boar and deer. Unfortunately, the modern-day threats have now shifted to the lack of strict laws to protect wildlife habitats in Japan. Laws and regulations potentially have significant influences on biodiversity conservation, and historical overviews are given later in this chapter.

1.4 Biodiversity in relation to ecosystem services and conservation

This book reviews two key attributes for national biodiversity conservation. Species richness is addressed by the number of indigenous species living in a specific area. The second attribute, species uniqueness, accounts for conservation value in an aggregated manner within a certain geographical range. This is illustrated by the terrestrial vertebrate species endemic to the Nansei Archipelago, in Amami Oshima, many of which are on the national species Red List. An alternative approach more recently summarised by Ingram *et al.* (2012) is the biodiversity-based ecosystem services. The review article points out that there is a wide range of services provided by the presence of biodiversity with large societal benefits. These "services" include providing food and timber, maintaining genetic resources, regulating environmental pollution and carbon sequestration, pollination, controlling pests and vectors, and cultural services represented by tourism, landscape aesthetics (e.g., Balvanera *et al.*, 2006; Elmqvist *et al.*, 2010; Haines-Young and Potschin, 2010; Atkinson *et al.*, 2012; Harrison *et al.*, 2014). Generalised or conceptual arguments on both biodiversity and ecosystem services are unfeasible (Ridder, 2008) and instead attention should be directed to the actual species populations or to specified communities that provide distinct services postulated to nominated aims. The following chapters present case studies on ecosystem services in specified regions or populations.

In the era of capitalism, the value of biodiversity is often undermined because there is no observable market value. General public often fail to recognise and appreciate ecosystem services that are dependent on biodiversity because there is no monetary revenue or profit made. A very prominent example of this is the promotion to clear natural forests to generate financial benefits in the short term, creating irreversible damage to the precious environmental assets where regeneration and recovery is now unfeasible. To offset these past failures, presently, there is an increased resolve in initiatives to advocate national conservation efforts. One of the most prominent of these is in the Nansei Archipelago. The studies on Amami Oshima classify wildlife into five groups according to the benefits and costs of services provided to that of the local people from each species, which will be further elucidated in Chapter 2.

A list of wild vegetable and mushroom species that people harvested was surveyed and published in Fukushima Prefecture in order to demonstrate the wide range of species people obtained benefits from in the presence of a diverse natural resources (Chapter 4). The broad variety of harvested vegetation in beech forests can be credited to contributions of the seasonal heavy snowfall which occurs in this region (Kitamura, 1998; Takahashi and Matsuura, 2011; Kawarasaki and Sugimura, 2012; Matsuura *et al.*, 2013). The third case study, presented in Chapter 5, conducted a statistical analysis in an attempt to assess the indirect relationship between biodiversity and human behaviour. These studies show the importance of provisions of services by the natural biodiversity to anthropologic culture. Ecotourism is one such service, where in Japan, the biodiversity-rich assortment of natural landscapes attract many people to experience them through eco-tours. Take and Saito (2011) examined many research papers on guided tours and their sites. The top five ecotourism attractions listed are two islands in the Nansei Archipelago, the Ogasawara Islands, a mountain range near the northern tip of Honshu and a peninsula in Hokkaido. UNESCO has designated four of these as World Natural Heritage sites, in addition to a set of four national parks. Almost all published papers on ecotourism have exclusively studied social aspects of ecotourism, such as its contribution to the local economy and visitors' behaviour and perception of local people. Independent of the aforementioned studies, satoyama activities that contribute to an increase in the diversity of plant species through shrub removal or thinning have been investigated to assess the relationship between species biodiversity and recreational activities. These activities aim to maintain appropriate forest conditions and quality for recreation and leisure to their visitors (e.g., Matsumura *et al.*, 2000; Yamazaki *et al.*, 2000). There are no studies that attempted to analyse the relation between recreational activities and species diversity. One reason is that biologists are not interested in studying social aspects of biodiversity, whereas social scientists are not very familiar with biology of plant and animals. Cross-disciplinary studies are rare and not encouraged. Another reason may be that academic societies are segregated by disciplines.

Contributions of biodiversity on environmental regulatory services involve complex ecological processes and are difficult to clarify and quantify. However, there are some significant notable examples. Fields for Japanese buckwheat production, where its commercial demands are high, are usually in areas where they are surrounded by forests and natural grasslands (Taki *et al.*, 2009). Seeding relies heavily on a great variety of pollinators, mostly honeybees and other small insects from these natural habitats. Maintenance of a high species diversity, which is conceptually vital for a more resilient ecosystem for changing environments compared to a community consisting of a few select species of honey bees, is crucial to the longevity and survival of this industry. Rice production is another industry, critical in Japan, where the rice paddy fields are reliant on the natural ecosystem and a high level of biodiversity. For the conservation and the protection of national endemic fauna, management efforts have been introduced. Moderating farming practices and management will reduce anthropological impacts on species diversity and subsequently boost benefits from environmental regulatory services. Nishikawa (2015) introduced various farming techniques to maintain species-rich ecosystems for the sake of protecting the populations of Japanese crested ibis introduced from China to Sado Island. Inagaki *et al.* (2015)

analysed management techniques to control a pest species, stink bug *Stenotus rubrovittatus*. No studies have been conducted in Japan on carbon sequestration services, another critical environmental regulating services approach. However, based on a number of studies conducted overseas, Yamashita *et al.* (2013) concluded that these services are heavily dependent on species composition and functional group diversity. Unfortunately, it is not currently possible to refer to these environmental regulating services further without additional regional and geographical analyses.

Within this book, I introduce research findings and personal experiences on biodiversity conservation and ecosystem services, such as endangered endemic species, harvesting wild plants and mushrooms and recreational activities as well as forestry (Figure 1.2) based on subdividing Japan into four regions (Figure 1.2). These regional boundaries are arbitrary, dependent on climatic conditions, high endemism, forest cover type, landscape use and human population density as criteria for the geographic division.

Kawarasaki and Sugimura (2012) gives an example of geographic analyses based on a generalised linear model, in which people in cooler climate regions (Figure 1.2 A)

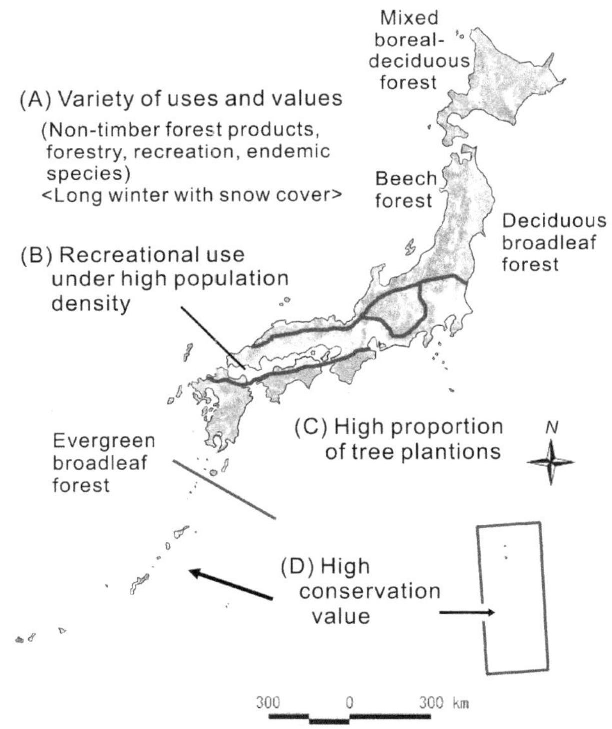

Figure 1.2 Geographical division based on regional variation in biodiversity and landscape use.

harvest mushrooms and wild vegetables more frequently than in warmer regions. The northern part of Japan has larger natural forests than the southwestern part, which benefit the harvesters who prefer natural forests to tree plantations with even-aged conifers (Matsuura *et al.*, 2013). Chapter 4 describes such activities in beech forests in addition to recreational fishing, although they were negatively affected by the nuclear power plant accident in 2011. An excellent review was presented by Saito (2006) where levels of dependence on wild plants and mushrooms were analysed in terms of quantity and variety of local cuisine using data collected from as many municipalities. To the south of this region is the Pacific Belt Zone where population density is very high. Outdoor recreational activities are most commonly observed in the satoyama areas of this region (Kawano, unpublished) and in some extremely crowded okuyama areas during the holiday seasons (Figure 1.2 B). Chapter 5 presents a case study that analysed recreational activities, population distribution and bird and plant species diversity in a semirural area (satochi) about 50 km north of Tokyo. The first sections of Chapter 6 focus on extensive tree plantations in the southwestern part of Japan (Figure 1.2 C). The region is characterised by low species diversity of birds, less popular nature tourism, and satoyama activities that are more concerned with forestry activities than Kanto region and Fukushima. The southernmost region, the Nansei Archipelago and the Ogasawara Islands, have the largest number of endangered species that are endemic to Japan (Figure 1.2 D).

1.5 Recent history of human impacts and nature conservation

In small alluvial plains along the coast and mountain basins, natural habitats have been altered drastically due to urban sprawl, land use changes associated with modern high-production agriculture and other various economic activities. As described in the "Introduction" section, coniferous trees were either planted or hardwood regenerated naturally after abandoning coppice management practices in the middle of the last century. Thus, the satoyama landscapes changed radically. In contrast, forested landscapes make up consistently about two-thirds of the land area, located in mountainous and hilly terrain. However, there were fundamental landscape changes also in okuyama after the Pacific War. Untouched okuyama forests were cleared extensively following the national policy as described below. We need to consider biodiversity conservation from the perspective of landscape changes both in satoyama and okuyama as well as government forest policies, protected area management and endangered species conservation.

1.5.1 *Shifting focus on forest management from economy to ecosystem services*

In 1964, Forestry Agency[(1-1)] enforced the Forestry Basic Act. The objectives of this act emphasised that forestry should contribute to economic growth and to an increase in the income of forestry workers. It specified that wood production was the most critical primary function of forests and that hazard mitigation was secondary, essentially ignoring all the other functions. As a government incentive to increase wood production, financial aid was stipulated in order to meet 100% of domestic demand for wood from Japan's public and private forests should be able to. This act also addressed that the national forests administered by the Forestry Agency must maintain a corporate and self-supporting accounting system, while it contributed to achieving the goal for wood production. In the '1970s, the demands of the act, which also encouraged clearing untouched forests in okuyama areas, brought about severe conflicts between conservationists and forestry agencies. One of the problems with this act was that it did not pay attention to regional differences in climate and local site conditions.

As a result of the national forestry policy, tree plantations that covered 22% of the entire forest area in 1951 increased to 44% in 2000. These tree plantations replaced many of the old natural growth forests in remote okuyama areas, where wood was chipped to produce paper. Thus, many arboreal species dependent on old growths must have suffered from severe degradation of their habitat conditions. Protests occurred across the country against such destruction of undisturbed ecosystems in the 1970s but almost all in vain. In particular, oppositions were strong to some of the clear-cuttings in the national forests which account for about 30% of the forested area in Japan. Hokkaido has the highest proportion of the national forests, where large areas of intact forests remained. The national forests across the southern and western parts of the country are considerably smaller. Now, only a few exceptions remain in Hokkaido, Tohoku and Yakushima Islands, all of which have been designated as UNESCO World Natural Heritage sites.

The great majority of tree plantations are even-aged and made of single species, about 80% of which consists of one of the following three species in 2017: sugi cedar *Cryptomeria japonica*, hinoki cypress *Chamaecyparis obtusa* and Japanese larch *Larix kaempferi* (Rin'ya-cho, 2018). The percentage of tree plantations compared to the entire forest area is high in the southwestern part of the country and lower in the north, 56% in Kyushu and Shikoku combined and 33% in Hokkaido and Tohoku, respectively. This high percentage corresponds to large habitat areas suitable for the cedar cultivation in southwestern Japan. The southernmost prefecture, Okinawa, has the lowest proportions of plantations at 11%, followed by three prefectures located along the Japan Sea. Fortunately, this is due to the warm subtropical climates of Okinawa, and conversely the other three prefectures are too cold with too much snowfall, both of which are unsuitable for cedar growth. The prefectures with cold climates have relatively warm and moist winds that flow over the Japan Sea in winter and hit against the mountains; thereby the area has one of the heaviest snowfalls in the world. Such regional variance is primarily on account of climate. The cedar can grow as an economically viable species across the majority of the country, if not too cold or warm; hence, it is favoured as a plantation species covering over 44% of the total plantation area. The cypress can be cultivated in a similar area to the cedar, with limited snowfall, and covers 26% of the total plantation area. Finally, the larch, adapted to cooler climates, has been planted in Hokkaido and northern

Honshu as well as on mountain slopes higher than 900 m above the sea level (Nagamitsu *et al.*, 2014). Government subsidies bore the cost of building forest roads, planting seedlings after clearing natural forests, weeding, thinning and other silvicultural practices. These practices continued until 2001, when the Forestry Agency rephrased the act as Forest and Forestry Basic Act, acknowledging that other functions and services by the forests are just significant as wood production. Chapters 2 and 4 report case studies that describe clearing of old-growth natural forests for chip wood without observation or consideration to biodiversity conservation and ecosystem services.

From 1964 to 2001, a period of 37 years passed before the Forestry Basic Act incorporated the word "forest" to the Forest and Forestry Basic Act. This act represented a fundamental shift from timber-oriented forestry policy to a policy emphasizing environmental aspects as well as forestry. The aim of the renaming act was to restore and maintain multifaceted forest functions, including nature conservation, human health and global warming. One of the influences of the new act is that Forestry Agency decided to encourage replacing conifer trees, which did not grow well due to unsuitable climate conditions, with broadleaf trees to restore some of the original forest ecosystems. Consequently, this policy brought about a decline in the proportion of tree plantations from 44 to 41%, in 2012, and has been maintained at this level since. Another key change was the top-down decision-making governance from the national government alone to role-sharing between local and national government. This change corroborated the difference in policy direction between the southwestern region, Kyushu and Shikoku, and the northern regions, Tohoku and Hokkaido. The southwestern regions advocate wood production, whereas in the northern regions plantations have largely been abandoned and replaced with natural forests. The major policy change has resulted in further difficulties in zoning. Previously, conifer plantations were considered suitable not only for wood production and land and water conservation but also for the other functions. Hence, land zoning for forest management was based on economic feasibility—this feasibility analysis considered government subsidies as incomes, which were virtually social costs. Now, zoning is more complicated, considering natural forests suitable for recreation and biodiversity, while regarding mature forests, either planted or naturally regenerated, more suitable for land and water conservation than younger stands (Sugimura and Howard, 2008).

There were two significant drivers for the shift in government focus from the wood production-oriented policy to a more environmentally sustainable policy. One was an increase in imported timber, which reduced domestic wood supply to 20% of the total, which occurred after the tariff on lumber was eliminated in 1964 during the period of high economic growth, when the wood supply increased rapidly (Figure 1.3). Since, imported wood outcompeted domestic wood and increased the share further until recent years, when the domestic supply gradually increased partially due to the increasing demand for biomass energy (Kuboyama, 2018). Thus, the government failed to achieve one of the goals, i.e., meeting 100% of domestic demand for wood, performing well below the target

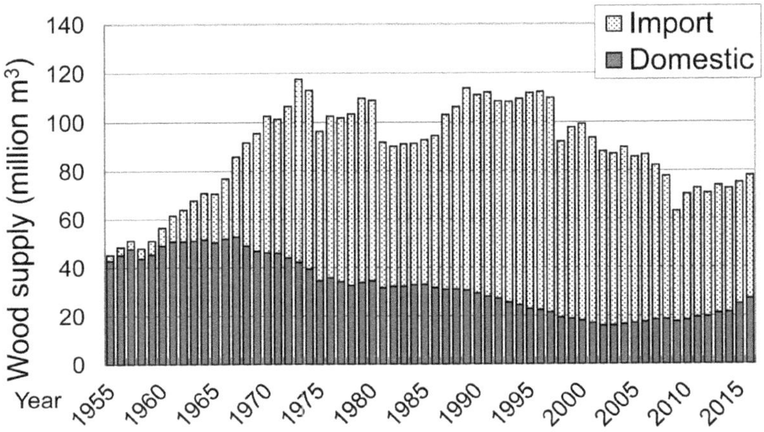

Figure 1.3 Changes in the volume of imported wood and domestic wood supply.

Source: Rin'ya-chō (1957–2018).

of the Forestry Basic Act of 1964. The other driver was the huge budget deficit that the national Forestry Agency caused, revealing that it was unable to maintain a self-sustaining accounting system. The agency was returning profit regularly until 1976, then turned into a persistent deficit (Matsuoka *et al.*, 2000). The deficit from the policy persisted for years to the final straw of ¥380 billion in 1998 (ca. 3.5 billion US$ at the exchange rate in 2019) (Nonaka, 2008), when it was considered to be as one of the most serious social issues in the country. Therefore, the federal Forestry Agency concluded that it was not economical to maintain wood production solely for commercial interests and thus should focus more on long-term asset management by emphasizing other forest functions. Nevertheless, the self-supporting accounting system lasted until 2013.

Fundamental changes to the laws were inevitable. Public opinion polls identified a drift in awareness, values and perspectives had changed, recognizing the importance of the forest as public goods rather than simply for wood production and government economic gains. This change in awareness must have acted as another impetus, even though the general public did not have sufficient knowledge about the forest functions. Random surveys opinion polls of approximately ten thousand members of the public are conducted annually. From 1976 to 2011, the poll included a supplementary survey on perception of forests contributing to human quality of life every four years. One of the questions attempted to identify which forest functions they thought were essential. From Figure 1.4, it is evident that there was a change in public perception and awareness by the respondents. In 1976 and 1980, wood production was identified as the second-highest percentage of importance next to hazard protection; subsequent surveys showed a decrease in its rank to the lowest from 1999 to 2007. Public awareness of how forests contribute to the quality of the water supply, which was ranked second, is reasonably limited. This is due to the complex nature

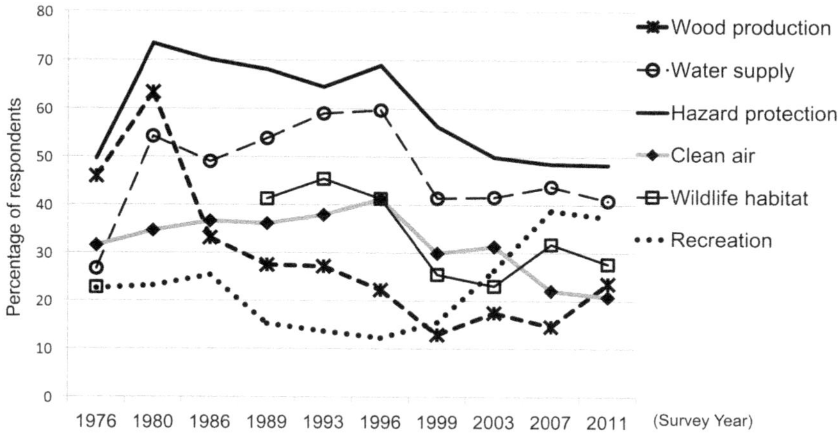

Figure 1.4 Percentage of respondents in polls. They chose forest functions that they thought were relatively important. Respondents could select up to three features.

and difficulty in quantifying the tangible effects. Trees deprive water of their ecosystem through evapotranspiration, while conversely forest soil can retain a considerably higher volume of water than farmland or urban soil. The landslides caused by heavy rains wash forest trees from time to time downhill, causing more damage than treeless land. However, it is not widely known that forests can accelerate disasters, and risk protection was ranked highest in public opinion polls. Given the lack of knowledge of the general public, it would make more sense to rank wood production higher than the polls indicate. Yet the Forestry Agency must have acknowledged that there was a clear discrepancy between the awareness of the general public and the agency.

1.5.2 *Protected areas and laws for biodiversity conservation*

Protected areas are one of the most effective tools for biodiversity conservation (Ara'ujo *et al.*, 2007; Mathur *et al.*, 2011). The Ministry of Health and Welfare governed these protected areas in Japan prior to the establishment of the Environment Agency (present-day the Ministry of the Environment) in 1971. The Ministry of Health and Welfare, focussed on human well-being and mental health rather than environmental protection. When environmental pollution, primarily of water and air, became a severe social issue, provisions and facilitation were promoted for the establishment of the Environment Agency.

It was in the early 1900s, the national parks in the United States were presented to the Japanese public for recreational activities in media outlets such as magazines. Meanwhile, government officials inspected Yosemite and Yellowstone National Parks, and subsequently, the Imperial Parliament submitted a proposal and filed a petition in 1911 regarding the establishment of the first national park in the country. The government enacted the law in 1931, and the first national park was established in 1934. The aim of the project was intended to implement the arrangements of the national parks in the United States to protect natural landscapes within Japan. However, Maruyama (1994) noted that the true purpose was to promote the local economy by managing parks with government subsidies for tourism. As such, the primary role addressed was to provide scenic beauty for visitors. Two-thirds of the 12 oldest parks were established before the Pacific War. They are either in Kyushu, southwestern Honshu or close to Tokyo, where a great majority of the urban population enjoy recreational activities at hot springs, golf courses or ski resorts while staying in large hotels with easy access to resort facilities, rather than nature-based recreation. The only exception is Chubu Sangaku (central mountain) NP, about 200 km away from Tokyo to the northwest, where many backpackers climb up mountains over 2500 m high. In contrast, the parks in the north and high-mountain regions, where population density is low, and natural forests cover relatively large fractions of the land area (Table 1.3), are generally more concerned with nature conservation than those in western Honshu and Kyushu.

Under the Ministry of Health and Welfare, the first three national parks were established in the Seto Inland Sea (Setonaikai) with its coastal areas, Unzen and Kirishima. The Seto Inland Sea is north of Shikoku and south of the westernmost part of Honshu. The most attractive element of the Setonaikai NP was supposed to be scenic beauty created by many islands in the sea. However, regrettably, the establishment of this park failed to prevent water pollution during the period of rapid economic growth in the 1960s and 1970s. It was before the establishment of the Environment Agency, which implemented robust measures to prevent water pollution in the 1970s. The government financed the development of heavy industry in selected areas centred on petrochemical complex along the

Table 1.3 Area of forests (km²), ratio of tree plantations and population density by region

Region[a]	*Forest*[b]	*Tree plantation*[c]	*Natural forests over 60 years old*	*Land area*	*Population density*
Hokkaido	55,425 (71%)	27%	56%	78,421	65
Tohoku	47,088 (70%)	41%	27%	66,891	134
Kanto	14,168 (44%)	45%	17%	32,426	1326
Chubu	47,387 (71%)	39%	25%	66,801	321
Kinki	21,859 (66%)	51%	11%	33,121	758
Chugoku	23,179 (73%)	41%	8%	31,922	233
Shikoku	13,993 (74%)	61%	6%	18,807	205
Kyushu	25,051 (76%)	58%	6%	33,004	334
Nansei Archipelago	2664 (57%)	18%	27%	4643	316
Total	250,814 (67%)	41%	26%	372,919	334

Notes

a The regions are arranged from north to southwest in the descending order.

b The figures in the parentheses are the ratio of forest cover to the total land area (the second column from the right).

c Tree plantation is expressed as the percentage of the forest area (the column on the left).

Source: Shinrin-ringyō-hakusho (Heisei 30-nendo) (1-2) for land statistics; Heisei 27-nen kokusei-chōsa 2015(1-3) for population

coastline, and the park law was not strong enough to stop them against severe water pollution. Unzen NP, which is in the northwest of Kyushu, is renowned for its hot springs, but the subject of biodiversity was neglected, as discussed further in Chapter 6. Unfortunately, biodiversity alone does not always appeal significantly to tourists, as demonstrated by Kirishima NP in southern Kyushu. Biodiversity-based tourism industry proliferated after the inclusion of Yakushima in 1964 to the national park.

Yakushima is an island to the south of Kyushu and has attracted visitors to old forests. UNESCO designated a part of the island as a World Natural Heritage site in 1993, which became an independent NP in 2012. The most popular destinations for ecotourism are the forests with ancient sugi cedars, whose estimated ages are over 3000 years. The island has Mt Miyanoura at an altitude of 1936 m, which is the highest in the southwestern part of Japan, hence distinctive vegetation change can be seen across altitudinal gradients. Yakushima has a long history of conflicts between nature conservation and forest production[1-4]. In 1952 the government proposed a plan to establish a national park on the island. However, in 1957, the national forests, which account for about 90% of the island's forests, began clear-cutting because old trees were no longer protected according to deregulation. Then in 1962, local individuals formed a conservation group to protest against logging, followed by a series of protest activities as well as the discovery of a sugi tree over 3000 years old in 1966, before the designation as the World Heritage.

There are 34 national parks as of the year 2020 under the jurisdiction of the Ministry of the Environment. In 1957, the Natural Parks Law was enacted, establishing a new system of natural parks that include prefectural parks and quasi-national parks. Among the three categories, the national park system is the most important in terms of area and quality of nature. The zoning system divides the parkland into special and normal zones. The special zone is further divided into four zones: special protection, first-class, second-class and third-class. The national parks cover about 6% of the land area in total, but limited areas are under strict protection. Only special protection and first-class special zones are comparable with the IUCN Category II, national park, where development activities are limited to small-scale tourism, covering in total 1.4% of the national land area ((IUCN is the abbreviation for the International Union for Conservation of Nature). In the second-class special zones, silvicultural activities, such as planting commercial trees and clear-cutting, are allowed through negotiations and compromises between the ministry and the landowners. The ministry grants permission for application of any land modification in the third-class special zones and has virtually no control in normal zones.

The quasi-national park has the same zoning system as the national park, while it is under the authority of individual prefectures. There are 56 parks, and the average area of a park is 252 km^2 (Kankyō-shō, 2019). The two most restrictive two zones cover 0.6% of the national land area. The main predicament of quasi-national parks is that usually they are not typically staffed with conservation specialists nor a park bureau, even though the office of civil engineering or that of agronomy and forestry is responsible for official processes, such as permitting a land use change within a park. The prefectural natural parks have, approximately, about the same total land area as the national parks, although the average size of each park is about a quarter of the national parks. In discussing the role of the prefectural parks, which may be fragmented but sizable as the sum, an examination of each park is necessary. Because the prefectural parks are often located near or next to a national and a quasi-national park, these parks can boost the total area of individual parks to provide a spatially larger essential habitat to protected species This is well-recognised from Species-Area curves (MacArthur and Wilson, 1967) indicating the area of a "Island," and hence spatial diversity, is essential for total biodiversity conservation. Prefectural parks have no special protection zones, and 10% belong to first-class special zones, hence small areas of strict protection relative to national parks (Sasaoka *et al.*, 1995). These natural parks in total cover nearly 15% of the land area in Japan, but this figure does not represent the performance of nature protection.

The park system has a severe drawback that is associated with land ownership in order to enforce preservation of biodiversity. The Ministry of the Environment owns only 0.3% of the parkland (Tanaka, 2012), thus the park system has severe drawbacks in negotiating agreements with landowners as to how to manage the land. Kankyō-shō (2001) reported that the national forests and national grasslands, managed by the Forestry Agency, covered a total of 12,202 km^2 (over 96%) of the national land in the national parks. As stated above, the Forestry Agency had been primarily concerned with wood production until 2001, even within the national parks except for the special protection zones and the first-class special zones. Thus, the agency encouraged local governments and private landowners to plant conifer trees after clearing natural forests of broadleaf trees sold for pulpwood in these zones. Local governments followed the national policy to obtain subsidies as described in Chapter 2. Areas under strict regulations for nature conservation are small even in the national parks. The encouragement of the Forestry Agency to expand plantations has therefore been a rigorous challenge for national parks to implement strict conservation measures. Table 1.4 indicates that park officers must negotiate conservation issues with various landowners. In the second-class special zone, there are no laws that specify permitted activities. Therefore, maintenance and preservation of this zone rely on a compromise within parks between the ministry and the landowners through negotiations, subjective judgements and examination of similar cases.

When the ministry plans to establish a new park, for instance, it has to obtain agreement with the owners as to the park boundaries and zoning inside the park, as discussed in Chapter 3. After the establishment, they have to monitor the activities of the owners and developers according to the park regulations. However, parks are severely understaffed; bureaucratic processing is the most time-consuming work for park rangers. Unlike the United States, there are no park biologists monitoring populations and analysing habitats for the species biodiversity to be protected.

Table 1.4 Area of the national parks by land ownership and the total land area of the regions (km^2)

Region	*National*	*Public*	*Private*	*Total*	*Land area*
Hokkaido	4543 (89%)	226 (4%)	324 (6%)	5092 (6%)	78,421
Tohoku	2652 (86%)	73 (2%)	376 (12%)	3250 (5%)	66,891
Kanto	1516 (50%)	338 (11%)	1161 (39%)	3016 (9%)	32,426
Chubu	2586 (59%)	952 (22%)	871 (20%)	4409 (7%)	66,801
Kinki	133 (10%)	175 (13%)	1073 (78%)	1381 (4%)	33,121
Chugoku	161 (27%)	99 (17%)	325 (56%)	584 (2%)	31,922
Shikoku	56 (14%)	39 (10%)	304 (76%)	399 (2%)	18,807
Kyushu	388 (23%)	350 (21%)	914 (55%)	1652 (2%)	33,004
Nansei Archipelago	600 (47%)	286 (22%)	329 (26%)	1282 (30%)	4643
Total	12,635 (60%)	2537 (12%)	5678 (27%)	20,917 (6%)	372,919

The numbers in parentheses (three columns on the left) show the ratio of the park lands in each category to the "Total" park area (the second column from the right) and the ratio of the total park lands to the "Land area."

Most of the "Public" lands belong to local municipalities, not to prefectures.

Since the 2011 Great Earthquake, the total of the three columns (National to Private) is less than the "Total" in Tohoku, as land ownership has not yet been determined over large areas.

Source: Kankyō-shō (2019).

Geographical locations also contribute to the variation in governance (Table 1.4). In the Kanto and Chubu regions where the proportion of nationally owned land is relatively large, the introduction of a park management system is straightforward due to the predisposition of the Forestry Agency to nature conservation initiatives. The prefectures and national forestry offices associated with the Shikoku and Kyushu regions are more inclined to support tree plantation management, as indicated by their high proportion of plantations at approximately 60% (Table 1.3). However, in the Nansei Archipelago, the proportion of parklands is considerably higher owing to the three recently established parks, two of which have specific directives for the conservation of endangered endemic species. In addition, this region is also in preparation for heritage site application to UNESCO.

The forest reserve system of the national forests was introduced in 1915, earlier than the national park system. Although it did not have a legislative basis, the Forestry Agency operates the system according to its internal regulations. This has enabled the preservation of large areas of nearly pristine forests, which were subsequently incorporated into national park systems at its core. Its role protecting the sizable areas of old growths formally ended in 1957, when the agency's role was declared to be transferred to the role of the national forests in nature conservation to the national parks. The Forestry Agency proposed policies to expand wood production capacity in order to meet increasing domestic demand (Ōuchi, 1987) and ignorance of nature conservation external to the core part of a national park lasted until 1989 when the agency made fundamental changes in forest policy as mentioned above. Currently, the forest reserve system has a total area of 977 thousand hectares subdivided into three categories: forest ecosystem conservation area, biocommunity protection forests and protection forests for rare species populations with a total of 31 locations and 701 thousand hectares allocated to the first category, the largest in terms of total area and area per location. This conservation area, located inside a national park, is very unusual but significant in the sense that the land manager and the law operator share the same goals.

During the early '80s, Japan was known as "the Japanese economic animal", during which it was very secure in its economic power. From 1986 to 1991, the so-called period of the bubble economy, had tremendous impact on the forests in mountainous rural areas. Forestry Agency has the conservation forest system, which limits the size of the clear-cut to 10 hectares or less, mainly for water resource management and natural hazard prevention such as floods and slope failures. Its total area is more than ten times larger than the forest reserve system above, and conservation forests for water resource and hazard prevention account for more than 90%. The agency recognised that wood production had failed to meet the domestic wood demand and due to the strong Japanese yen against foreign currencies, the supply and demand equilibrium would be difficult to address. Conversely, almost the whole society was keen on expenditures for leisure activities at that time, and thus politicians decided that forests should be replaced by recreational spaces, such as golf courses and ski slopes. Subsequently, the Act on Development of Comprehensive Resort Areas in cooperation with National Land Agency and Ministry of Agriculture, Forestry and Fisheries was passed by the then Ministry of Economy, Trade and Industry. At this time, almost all municipalities, with vast areas of forests, tried to revitalise the local economy by attracting resort guests. Forestry Agency has facilitated the conversion from conservation forest to normal forest to enable clearance of large areas of forest. In the bubble economy, these short-sighted misconceptions continued until the "bubble burst" in 1991.

From 1989 to 1990, the Japanese society came to a realisation that environmental issues were gaining momentum on a global scale (Sugimura, 1990). Japan decided to join the Convention on Biological Diversity at the 1992 United Nations Conference on Environment and Development, aiming to be a leader in problem-solving for these environmental issues. Although in 1992, delegates probably did not distinguish the fundamental differences between global environmental issues, such as climatic change and species extinction, and local problems of environmental pollution, Japan was proud to be the world's most advanced country in pollution control technology, unrelated to biodiversity conservation. However, this event was the turning point for nationally addressing local conservation issues, and the Ministry of the Environment formulated the National Biodiversity Strategy in 1995. The ministry also endorsed laws directly related to biodiversity after this UN conference, such as Act on Conservation of Endangered Species of Wild Fauna and Flora (1992); and then years later, the Invasive Alien Species Act (2004) and the Basic Act on Biodiversity (2008) followed. The term "biodiversity" appeared for the first time in other laws, such as Natural Park Act amended in 2002 and Forest and Forestry Basic Act (2001). The Environmental Assessment Act could not be implemented for many years until 1997, due to opposition from the other ministries. As a result, over the last ten years, the ministry was able to establish additional six new national parks, and expanded an already existing park. Five of these have a larger proportion of designated special protection zones than previously allocated. The Yambaru and Amami-gunto national parks are the first national parks that were established to protect many unique endemic species rather than the emphasis placed on scenic beauty and tourism.

Supplementary notes

1-1) The Forestry Agency administers forests and natural grasslands, the latter sometimes maintained by controlled burns for grazing, and the potential natural vegetation is forest. The statistics in the text include land that has lost trees after logging and land that has no trees due to fires or natural disasters. In this document, the term "forested area" includes land with and without trees. Each prefecture has a forestry department in charge of private forests and local public forests, while receiving grants from the national Forestry Agency.

1-2) Available at: https://www.rinya.maff.go.jp/j/kikaku/hakusyo/index.html (accessed 30.07.19).

1-3) Available at: https://www.stat.go.jp/data/kokusei/2015/kekka.html (accessed 30.07.19).

1-4) *Yakushima sangakubu o chūshin toshita rekishi nenpyō* (Kankyō-shō). Available at: https://www.env.go.jp/park/yakushima/ywhcc/wh/arikata/1/161225-4-2-1.pdf (accessed 05.05.18).

2 Amami Oshima

A treasure island of unique species impacted by logging during the 20th century

Introduction

In October 1984, Prince Phillip, Duke of Edinburgh and the then president of the World Wildlife Fund (WWF), made a campaign trip to Japan to increase the amount of funding for conservation efforts there. During his stay, he visited a small island in the southern part of the country, Amami Oshima (hereafter "Amami" or "the island"). He had a great enthusiasm for the preservation of the unique fauna and flora on the island, in particular the Amami rabbit *Pentalagus furnessi*, which scientists thought lived in dense forests. His visit made a significant impression on many local people, leaving the memory in their mind for quite a while. Before he left Japan, the Duke asked even the Emperor of Japan (Ten-nou) not to let any of the island's forest be cut down for the sake of this valuable rabbit species. However, extensive clear-cutting of the island's evergreen subtropical forests continued for more than a decade after his visit, with little attention paid to the habitat quality for the rare species living in the island. Amami became well known overseas, but local conservation groups had little control over the logging.

Nevertheless, during the 1980s and 1990s an increasing number of people, living on the island and from outside, began conservation activities, including the monitoring of wildlife population. They protested not only against logging but other activities also, such as road construction and hunting, which were taking place in relatively untouched forests. However, forest managers recognised the forests primarily as tree farms for woodchip production. At the same time, they maintained that there was no scientific evidence to demonstrate that clear-fell logging would have negative impacts on any of the species endemic to Amami and nearby islands. On the other hand, there was also no economic analysis that showed forest logging was the most productive way of exploiting the resources. In order to find reasonable solutions to these problems, I initiated a series of studies which collected quantitative data on some wildlife species, to explore the flexibility of the logging system with a comprehensive and integrated approach to mitigate the conflict. The studies asked two basic underlying questions about the conflict between forestry and species preservation. First, what were the effects of logging on the populations of local wildlife species? Second, was it possible to alter the current forestry practice in order to protect some rare species from any adverse effects of clear-cutting?

At the same time, however, some wildlife species were not only impacting the human society through damage to agriculture but also providing benefits to the society in terms of non-consumptive recreation and aesthetic values. These interactions between humans and wildlife may be woven together in an intricate pattern through the local human communities and forest ecosystems. In order to develop a reasonable forest policy for wildlife conservation in this situation, it would be necessary to study wildlife populations to examine the effects of human disturbances. Also, we need to identify what factors in the human society play significant roles in the human-wildlife relationships.

2.1 Natural geography and society

2.1.1 *Location and natural geography*

Amami Oshima is located at latitude 28 degrees north and longitude 129 degrees east in the central Nansei Archipelago (Figure 2.1). It is far from any large cities, i.e., it is 370 km away from Kagoshima City, the capital of Kagoshima Prefecture to which Amami belongs, and the main departure point for air and sea travel to the island. It is a little closer to Naha, the largest city in the archipelago and the

Figure 2.1 Locations of Amami Oshima and other islands in the Nansei Archipelago.

capital of Okinawa Prefecture, but transportation is much less convenient. As an administrative unit, the island belongs to the Amami Island Group County, which has five major islands and Amami City (formerly Naze) serves as its capital[(2-1)]. Amami, the largest of the five islands (area 719 km^2), has 56% of the total population. All the statistical information used here is from the official data book[(2-2)] that Kagoshima Prefecture publishes each year, if not mentioned otherwise.

The island is the second largest in the Nansei Archipelago, and it has many small alluvial plains used for agriculture and human habitation. Most of the plains are topographically isolated from each other by steep slopes along the coast; this used to hinder frequent travel between local communities. The rest of the island is mountainous, generally with steep slopes near the coast and covered by broadleaf evergreen trees. Thus, when we walked along small streams looking for rabbit pellets, we usually encountered a waterfall so had to be careful crossing them hoping not to meet a local venomous snake, the habu *Protobothrops flavoviridis.* Above such slopes, the terrain generally becomes gentler towards the ridge, around 300–500 m high. The highest peak is Mt. Yuwan, 694 m above sea level. It has a shrine at the top, but local people seldom go there, probably because it is too hot most of the year.

The climate is often described as subtropical, having coral reefs and the second largest mangrove forest in the country. The sea surface temperature around the island is above 20 degree Celsius all the year round, according to Dai-jū kanku kaijō hoan honbu kaiyō jōhō (2018). The annual average precipitation for 30 years from 1981 to 2010 was 2800 mm[(2-3)] and a local weather station on the island reported that the average total hours of sunshine a year was 1360 hours for the same period. This is the lowest in the country, which is 3.7 hours a day on the local average. Rainfall varies significantly by season. From December to early March, cold northwest winds frequently prevail, blowing from China across the East China Sea; they occasionally produce orographic rainfalls accompanied by wind blowing up the steep slopes. This season was the best to do fieldwork in the forests due to a much lower chance of meeting the habu than the other seasons.

In the mid-1980s, I often saw vehicles parked on forest roads on weekends. Most probably their owners were collecting orchids and other rare plant species, which they could sell at a high price in the black market. We do not see them very often now, most probably because the species are almost extinct in the wild rather than due to growing conservation awareness. Towards the end of winter, warm spring winds blow more often from the south, and the Amami locals start watching out for the habu. Forests are full of bird songs and the new leaves of evergreen oaks. The season changes from relatively cool to hot and rainy. It is generally in the middle of May when cicadas, Cicadoidea, start their loud shrilling and most of the birds stop singing. Towards the end of the rainy season, sudden falls of rain become heavier, then typhoons come close to the island one after another. Autumn is not special until the oaks produce acorns in November, providing abundant food for many species of wildlife and insects. However, acorn production is not predictable because of large annual variations.

2.1.2 *Amami's economy and society*

Islands tend to form a unit of local administration and economy. Amami provides an excellent example of local Japanese society, in the sense that its population has been declining for several decades and that its economy is dependent on government subsidies. Financial dependence of the local government on the central government is common across the country (Sakamoto, 1994). Only metropolitan cities and prefectures with a large population can be financially independent. There are two common mechanisms by which local government receive financial aid from the central government, namely chihō kōhuzei (local allocation tax) and kokko-shishutsukin (treasury expenditure). Remote islands receive an extra subsidy on top of these two, and the extra budget for Amami is especially large, resulting in a massive total subsidy.

This chapter deals with field studies on the effects of logging on the wildlife species in Amami that I began in 1985. Consequently, statistics in this section are mainly based on the year 1985. However, sometimes it is important to see how conditions have altered. In such comparisons, the most recent data are also given with older data.

The island's population declined from 84,798 in 1985 to 61,256 in 2015 (28%) but fell by 41% from 1955 to 2015. Urbanisation also proceeded, and more than half of the population now lives in the central area of Amami City, where most tourists stay. Even the population of this city has decreased since 1990. Population decline in Japan occurs nowadays not only on remote islands but also from Hokkaido to Kyushu—except for metropolitan areas and the core city of a region, such as the capital of a prefecture. For example, the population of Kagoshima City increased by 70,000 to 600,000 between 1985 and 2015, mainly due to immigration from rural areas in the prefecture.

When Amami was self-sufficient in food, the alluvial plains were insufficient, so farmers had to cultivate hillsides or reclaim land by draining it. As the number of farmers declined sharply, terraced fields on slopes were abandoned to be replaced by naturally regenerated forests or converted to orchards. Farmland covered only 2.7% of the land in 1985, which was a 33.6% decrease from 1960. The agricultural labour force in Amami decreased from 4.9% of the population in 1985 to 3.2% in 2015. This is a much smaller percentage than that of the rural area in Kagoshima Prefecture, and even smaller than that of the nation as a whole—9.8% in 1985 and 3.8% in 2015. Of the island's agricultural commodities, sugarcane fields cover the largest land area. The national government encouraged the growth of sugarcane as a cash crop, so from 1966 farmers gave up growing subsistence crops, such as rice and sweet potatoes, making a monoculture system dominant on the island (Hasumi, 1981). Local sugar is not competitive against sugar imported from developing countries, so the government purchases the local product at a higher price than the wholesale price.

There are a few disadvantageous conditions for local agriculture. One is the island's infertile non-volcanic soil, which has been leached by spontaneous heavy rains and continuous precipitation throughout the year. The government attempted to improve soil fertility by bringing more fertile soil into farmland. However, some local people claimed that heavy rains washed away a considerable amount of the transported soil to cover and killed coral reefs, causing serious harm to local fisheries.

Another problem is that pests have a longer breeding season, so they can breed more times a year and multiply faster than in areas north of Amami. For instance, one species of flower fly (*Dacus cucurbitae*) had caused serious damage to vegetable production and required a very intensive and expensive government pest control project that took several years for eradication. Farming also has a high risk of snake bite, because venomous snakes are more in farm land than in forests and requires hard work in hot and humid weather during the long summer as well as the cool and rainy winter. In 1985, 88% of the farming households were engaged in another job as the primary source of income (Sugimura, 1987). As a result, the people of Amami depend heavily on the cargo ships coming from Kagoshima City for their food.

Amami's secondary industry employed the largest number of people. Its textile industry employed the largest number of people in 1985. This used to be the island's largest self-supporting, productive industry producing the island's most important market wares, Oshima tsumugi—a pongee fabric with a special texture for making Japanese kimono. Textile manufacture was considerably more profitable than any of the island's industries during the 1970s. However, it struggled with serious depression in the mid-1980s (Sugimura, 1988). At that time, the manufacturing industry alone employed about the same number of people as the island's whole primary industry. However, the number was about one-third of the agricultural labour force in 2016. It is still valuable as part of Amami's unique traditional culture, producing souvenirs as well as tourist attractions. However, there is no hope of reinvigorating it as one of the island's major industries. Japanese people do not wear kimono as often as before, while Oshima tsumugi imported from Korea is much cheaper and popular. Given this situation, how has the island's economy been sustained?

The national government has maintained a special aid project for the county since 1954 (The Amami Archipelago Promotion and Development Project: AMASHIN). There was a good reason for the government to start such a project for economic reconstruction and development, the financial contributions for which have been enormous given the county's economic size.

After the Pacific War ended in 1945, the United States occupied and governed the island for eight years, during which time many people suffered from food shortages and starved nearly to death. This extra subsidy seems to be a kind of compensation for that challenging period. As a result of the huge amounts of financial aid, civil engineering and construction became the most critical sector in the island's secondary industry in terms of net product, producing about half of the industry's output. The government budget for the project in the fiscal year of 1985[(2-4)] was ¥43,833 million, excluding the amount provided by local municipal governments. In total, all financial aid from the national and prefectural governments amounted to ¥84,025 million—44.5% of the total income of Amami Oshima County (Ibid). In 2016, this had increased to ¥102,415 million (about US$966 million), despite the declining population. The total population was then about 110,000, so the total amount of financial aid was equivalent to ¥930 thousand per person. This was about half the average national gross domestic product per capita. Significantly, of the total number of people employed in the county at 2016, 69% were in civil engineering, as illustrated later, these government subsidies significantly influenced forest use and biodiversity conservation.

2.2 Wildlife and their interactions with local society

2.2.1 *Biogeographical history*

During the glacial periods, although it is difficult to estimate the duration, the Tokara and Kerama Straits (Figure 2.1) were deep enough for the central part of the Nansei Archipelago to be separated from Taiwan and Kyushu. Ota (1998) suggested that the islands have been separated since Early Pleistocene, roughly about 1.5 million years ago, based on geological and biological studies. Palaeontologists found a large number of fossils of this period from the earth layers, such as elephants, deer, wild boar, rats and turtles (Kizaki and Ohshiro, 1980). The land area of the archipelago was larger than now due to the lower sea level during the ice age, and large animals were able to live there. Also, these layers left fossils of the long-haired rat *Diplothrix legata* and the spiny rats *Tokudaia* spp. that are endemic to the central Nansei Archipelago. The isolation after the sea level rise has resulted in the Archipelago having many endemic species of terrestrial animals (Ikehara, 1980; Ota, 1998; Masaoki, 2009). The fauna and the flora are distinctive not only by the number of endemics but also because some relic species have their closest phylogenetic relatives in Southeast Asia or Tibet (Kizaki and Ohshiro, 1980). In contrast, the biota of the other regions in Japan has a lower degree of endemism and in general species are more closely associated with that of the Asian relatives (Ibid). The Amami rabbit *P. furnessi*, which lives only on two islands, Amami and Tokuno, is a good example. It retains primitive characteristics, such as short limbs and ears, small eyes and a prolonged snout relative to most of the other species of lagomorphs (Photo 2.1). Fossils of such relic species are found throughout the world. However, their current distribution is quite limited at present: the volcano rabbit *Romerolagus diazi* in Mexico, the rock hare *Pronolagus crassicaudatus* in the eastern and southern parts of Africa and the Amami rabbit in Japan (Chapman and Flux, 1990).

Table 2.1 gives a list of endemic terrestrial vertebrate species and subspecies that are on Amami and on the Japanese Red List. Particularly noted is a large number, 20 species of reptiles and amphibians, reflecting a long period of geological isolation from the continent and their poor dispersal ability (Ota, 1998).

Photo 2.1 Amami rabbit *Pentalagus furnessi*.

(Photograph courtesy of Y. Takatsuki)

The number of endemic reptiles of the Nansei Archipelago is twice as many as the rest of Japan, suggesting a very high level of endemism despite the small land area. One-third of the species in the table have an even smaller distribution, being limited to areas to the north of Okinawa Island. Also, a very high population density in local forests could be inferred from the large number and various kinds of snakes one may encounter swimming in small streams in the late afternoon on a hot summer day. Such a scene is a partial representation of the rich species diversity of the island. Most of the snakes are nocturnal, while diurnal non-venomous species, the Ryukyu green snake *Cyclophiops semicarinatus* was abundant in 1985. Sometime later a large number of small Indian mongoose *Herpestes auropunctatus* invaded the central part of the island and the green snake was scarce in the 2010s. Another species that also became scarce was sword-tailed newt *Cynops ensicauda*, which used to be seen warming themselves on forest roads on a sunny winter day. The next chapter describes the effects of the mongoose invasion on some of the island's birds and mammals.

All the species in Table 2.1 live in forest-dominated landscapes. Mammals are nocturnal and more challenging to find than birds, so mammalogists have not succeeded in producing research outputs which forest management could adopt

Table 2.1 Terrestrial vertebrates on Amami Oshima that are endemic to the Nansei Archipelago (subspecies of particular concern are also included)

Scientific name	*Common name*	*Distribution*	*Red List status*[a]
Mammals			
Pentalagus furnessi	Amami rabbit	AO, TS	EN
Diplothrix legata	Long-haired rat	AO, TS, OH	EN
Tokudaia osimensis	Amami spiny rat	AO	EN
Crocidura watasei	Watase's shrew	AG, OG	NT
Crocidura orii	Orii's shrew	AO, TS	EN
Miniopterus fuscus	Ryukyu bent-winged bat	NS south of TR	EN
Myotis yanbarensis	Yanbaru whiskered bat	AO, TS, OH	CR
Murina ryukyuana	Ryukyu tube-nosed bat	AO, TS, OH	EN
Birds			
Garrulus lidthi	Purple jay	AO	–
Scolopax mira	Amami woodcock	AG	VU
Erithacus komadori	Ryukyu robin	NS, DG	VU
Zoothera dauma amami	Amami thrush	AO	VU
Dendrocopos leucotos owstoni	Owston white-backed woodpecker	AO	VU
Reptiles			
Japrula polygonata	Tree-climbing lizard	NS south of TR	–
Tokydromus smaragdinus	Green kanahebi	TR, AG, OG, MG	–
Ateuchosaurus pellopleuru	Dark-edged hime lizard	NS	–
Plestiodon barbouri	Barbour lizard	AG, OG	VU
Plestiodon marginatus	Oshima lizard	TR, AG, OG	–
Achalinus werneri	Amami Takachiho	AG, OG	NT
Cyclophiops semicarinatus	Ryukyu green snake	TR, AG, OG	–
Dinodon semicarinatus	Akamata	AG, OG	–
Amphiesma pryeri	Glass hibah	NS south of TR	–
Calliophis japonicus	Hyan	AG, OG	NT
Trimeresurus okinavensis	Hime habu	AG, OG	–
Trimeresurus flavoviridis	Habu	AG, OG	–
Amphibians			
Hyla hallowelli	Hallowel green frog	AG, OG	–
Rana okinavana	Riukiu red frog	AG, OG	–
Odorrana splendida	Amami Ishikawa frog	AO	EN
Odorrana amamiensis	Amami tip-nosed frog	AO, TS	VU
Babina subaspera	Otton frog	AO	EN
Rhacophorus viridis	Amami green frog	AG, OG	–
Echinotriton andersoni	Anderson's crocodile newt	AO, TS, OG	VU
Cynops ensicauda	sword-tailed newt	AG, OG	NT

Abbreviations: AG, Amami Island Goup; AO, Amami Oshima; CR, critically endangered; DG, Danjo Insular Group; EN, endangered; MG, Miyako Insular Group; NS, Nansei Archipelago; NT, near threatened; –, not on the list; OG, Okinawa Insular Group; OH, Okinawa Island; TS, Tokuno Island; TR, Tokara Insular Group; VU, vulnerable.

a The status is based on the national Red List.

Source: Sugimura (1987); Kankyō-shō (2018).

for species conservation on Amami. For instance, biologists have devoted far more efforts to the Amami rabbit than any other species. However, there are quantitative studies neither on its food habits nor population dynamics. It seems that acorns produced in autumn is one of its major food items and may have a significant effect on rabbit reproduction (Sugimura, 2000). Another species of mammal, the long-haired rat is an arboreal species that nests in tree cavities like a squirrel. It is the largest rat found in Japan, with a body length and tail length measured at 267 and 326 mm, respectively (Photo 2.2). In contrast, Imaizumi (1960) reported the body size of black rats *Rattus rattus* ranged from 150 to 235 mm with a tail length ranging from 150 to 258 mm. Owing to the long-haired rat's arboreal habits that require a tree cavity of a certain size, Sugimura (2016) roughly estimated that there should be trees large enough to have a diameter at breast height (DBH) over 30 cm in naturally regenerated forests more than 80 years old. However, extensive clear-cutting had occurred by 1985 for pulpwood production as described later. The area of clear-felled sites has decreased significantly for the past two decades, and the rats are observed more often in recent years than in the 1980s. Another unique species of rat, *Tokudaia osimensis*, also lives in forests. Biologists now classify this species as one of three closely related species based on a chromosome analysis, although it was previously considered a single species (Nakamura *et al.*, 2007). The other two species live on the Tokuno and Okinawa Islands. The morphological uniqueness of this species is its spines, being modified hairs, but we do not know their function. There are so many predatory snakes on the island that the rats must have some defence mechanisms and/or high reproduction rates. Dr. S. Hattori (a herpetologist at Amami Laboratory, the Institute of Medical Science, the University of Tokyo) told me that the rat could jump to escape from snakes and that it has high immunity to the habu venom. A dead female rat was found to have contained six foetuses, indicating a potentially high reproductive rate.

Photo 2.2 Ryukyu long-haired rat *Diplothrix legata*.

The national Red List includes eight mammal species living on Amami (Table 2.1). Many species in the table have phylogenetic relatives in Southeast Asia (Ikehara, 1980). The two species of rats have their closest relatives in Sumatra and Celebes. Little is known as to the ecology of the two shrews and three bat species. Not many Japanese universities and research institutes have wildlife conservation laboratories, and Amami is only one of the many potential research sites. They have to find sufficient funding for long-distance travel, while no scientist from the relatively close Kyushu and Okinawa regions went to the island until the mid-2010s. Also, it is difficult to efficiently conduct fieldwork on nocturnal and inconspicuous mammals (especially for peer-reviewed papers) when their populations might have declined to the brink of extinction during the periods of extensive clear-cutting and mongoose invasion.

There are not many endemic species of birds on Amami. However, taxonomists classified most of the indigenous subspecies as endemic to the Nansei Archipelago or the Amami Island group. Some of them have territory songs quite distinct from those subspecies breeding to the north of Amami. The Amami thrush *Zoothera dauma amami* is a good example. Its song has several syllables and sounds like the melody of a typical thrush species, while the other species living in Kyushu and further north have songs with one or two long syllables. Some ornithologists argue that these two forms should be two separate species, which will probably significantly increase the amount of effort to protect this thrush. Even so, they must have been isolated from each other reproductively for an extended period due to the geographical distance and thus maintained populations as if they are separate species. Other species with a small distribution range are the purple jay *Garrulus lidthi* and Amami woodcock *Scolopax mira*. No species is phylogenetically related to the jay, while the woodcock has a related species in the north[2-5].

Table 2.2 lists the vascular plant species found on Amami and endemic to the Nansei Archipelago, and many are on the Red List. The table includes 28 families, covering a variety of taxonomic groups. Orchids (Orchidaceae) suffer from the highest harvest pressure, but they are not endemic to the Nansei Archipelago. Their distribution generally extends to the south, so that they would be rare in Japan. One problem is that no well-designed study has clarified the characteristics of potential habitats. Therefore, it is evident that conservationists are not ready to propose effective protection strategies. Hayao Ohno, a prominent local plant specialist, told me that clear-cutting forestry had the most significant impacts on endangered plant species, adding to that of human harvest pressure. On the other hand, Dr. T. Yoneda at Kagoshima University indicated that forest logging is sometimes beneficial for edge-dwelling species, some of which are on the Red List. Studies conducted so far are too few to determine the effects of logging on plant species.

There used to be a considerable number of plant gatherers on Amami, and they sold unique plants for high prices in the black market. One time in 1986, my assistant found a kind of conk fungus in a forest, *Ganoderma applanatum*, which was thought to be effective for cancer treatment. He started breaking off

Table 2.2 Vascular plant species found on Amami which are endemic to the Nansei Archipelago and their Red List status

Species	*Family*	*Related plant in English*	*Red List status*
Ilex dimorphophylla	Aquifoliaceae	Holly	CR
Ilex macrocarpa	Aquifoliaceae	Holly	CR
Arisaema heterocephalum	Aracea	Jack-in-the-pulpit	EN
Aristolochia liukiuensis	Aristolochiaceae	Birthwort	–
Asarum celsum	Aristolochiaceae	Birthwort	EN
Asarum fudsinoi	Aristolochiaceae	Birthwort	VU
Asarum gusk	Aristolochiaceae	Birthwort	CR
Asarum lutchuense	Aristolochiaceae	Birthwort	EN
Asarum trinacriforme	Aristolochiaceae	Birthwort	EN
Balanophora yuwanensis	Balanophoraceae		–
Aster asgrayi	Compositae	Goldenrod	EN
Aster miyagii	Compositae	Goldenrod	VU
Solenogyne mikadoi	Compositae	Goldenrod	VU
Trichosanthes miyagii	Cucurbitaceae	Melon	–
Carex sakonis	Cyperaceae	Sedge	NT
Rhododendron scabrum	Ericaceae	Rhododendron	VU
Vaccinium amamianum	Ericaceae	Azalea	CR
Quercus miyagii	Fagaceae	Beech	–
Aristida takeoi	Gramineae	Needlegrass	CR
Salvia pygmaea	Labiatae	Salvia	EN
Scutellaria rubropunctata	Labiatae	Mint	–
Cinnamomum doederleinii	Lauraceae	Camphor tree	–
Lilium alexandrae	Liliaceae	Lily	CR
Ligustrum liukiuense	Oleaceae	Olive	–
Oxalis amamiana	Oxalidaceae	Wood-sorrel	CR
Lysimachia liukiuensis	Primulaceae	Primrose	CR
Rubus amamianus	Rosaceae	Raspberry	CR
Zanthoxylum amamiense	Rutaceae	Citrus	–
Cardiandra amamiohsimensis	Saxifragaceae	Saxifrage	EN
Deutzia naseana	Saxifragaceae	Saxifrage	CR
Gratiola fluviatillis	Scrophulariaceae	Figwort	EN
Veronicastrum liukiuense	Scrophulariaceae	Figwort	CR
Eurya osimensis	Theaceae	Eurya	NT
Elatostema oshimense	Urticaceae	Nettle	CR
Nanocnide lobata	Urticaceae	Nettle	–
Callicarpa oshimensis	Verbenaceae	Beautyberry	VU
Viola amamiana	Violaceae	Violet	CR
Diplazium amamianum	Athyriaceae	Fern	–
Polystichum obae	Dryopteridaceae	Fern	CR
Sphenomeris minutula	Lindsaeaceae	Fern	CR
Goniophlebium amamianum	Polypodiaceae	Fern	CR

Abbreviations: CR, critically endangered; EN, endangered; VU, vulnerable; NT, near threatened; –, not on the list.

Source: Sugimura (1987); KankyRLINK (2018); Amami Yasei Seibutsu Hogo-sentā(2-5); Tsukuba Jikken Shokubutsuen(2-6); Kagoshima-ken(2-7).

some pieces. I told him not to take it all, but he said that we should do so before somebody else found it. Probably, this was the typical strategy for collecting wild plants, which reflects the "tragedy of commons." Collectors may not have cared that such rare plants were seldom found outside the island, and they did not feel very guilty about eliminating the species populations. Such circumstances unfavourable for rare species appear to have since changed for better. Islanders have become aware that the biodiversity of Amami has globally unique values. Local people possibly came to recognise the importance of biodiversity conservation when the Ministry of the Environment started working towards UNESCO World Natural Heritage registration in the early 2000s.

2.2.2 *Human-wildlife interactions*

The wildlife species on Amami can be categorised based on their conservation value and costs to society. On the one hand, there are endemic species and subspecies, and on the other, those that act as pests or are dangerous to humans. However, some species can be both to varying degrees. Table 2.3 illustrates such relationships, classifying species into a range of benefits and costs with five groupings and describing the positive and negative aspects of each species. Government agronomists and forestry managers in Amami used to see wildlife species primarily as a nuisance and not worthy of preservation. They almost ignored the high levels of endemism because those rare species were common throughout the island. In contrast, conservationists did not pay much attention to the economic costs native animals can have on agriculture. The following sections examine and consider their various costs and benefits.

2.2.2.1 *Species with positive values*

The first group (A1 in Table 2.3) has almost entirely positive aspects, such as endemism, conservation values, aesthetic and economic values. It is far larger than the other groups, suggesting that wildlife species are generally beneficial to the local society. In contrast, however, the most commonly observed mammals in Kyushu—the wild boar and Japanese deer—are harmful to agriculture and forestry, as mentioned in Chapter 6. Five island species are on the national Red List. Considering taxonomic uniqueness and range of distribution, the purple jay and spiny rat are the most valuable. The jay is found only in Amami and a few small neighbouring islands, having beautiful bright blue and reddish-brown feathers as well as a noisy call, characteristics typical of the crow family (Photo 2.3), and with the Amami rabbit it is representative of the island's unique biota. The Ryukyu robin *Erithacus komadori* has the widest distribution of the endangered species, and its population size is probably the largest of the island species on the Red List. It is distinguished by its beautiful songs, lovely looks and extreme curiosity (Photo 2.4). During the feudal period[2-8] before 1868, Satsuma-han (the local government area that geographically overlaps with Kagoshima Prefecture)

Table 2.3 Positive and negative values of wildlife species in Amami perceived in the middle of 1980s

Species	*Status*	*Type of values, monetary costs and benefits (¥thousand/year)*	*Group*
Long-haired rat	NM, RL	Endemic to Amami Oshima	(A1)
Amami spiny rat	NM, RL	Endemic genus	(A1)
Ryukyu robin	NM, RL	Beautiful song call	(A1)
Purple jay	NM	Endemic to Amami Oshima	(A1)
Amami thrush	NM	Very small population	(A1)
Amami woodcock	RL		(A1)
Oriental white-eye		Pet	(A1)
Insectivorous birds (eight common species other than the three NM avian species above)		Pest control	(A1)
Amami rabbit	SNM, RL	Endemic genus; econ. cost: –2122 (fr)	(A2)
Owston white-backed woodpecker	NM, RL	Insectivorous; econ. cost: –348 (fr)	(A2)
Ryukyu wild boar	Game	Tax revenue[a]: $+4.8 \times 10^3$ Meat[b]: $+30 \times 10^3$ Econ. cost: –28,520 (ag)	(B)
Formosan green pigeon		Econ. cost: –16,460 (ag)	(B)
Habu		Commercial harvest Venomous bites	(C1)
Brown-eared bulbul		Econ. cost: –30,341 (ag)	(C2)
Jungle crow		Econ. cost: –14,810 (ag)	(C2)
Eastern turtle dove		Agricultural damage reported on other islands	(C2)
House rat		Econ. cost: –1199 (ag)	(C2)

Status: The Agency for Cultural Affairs protect special natural monument (SNM) and natural monument (NM) species from catching; RL denotes that it is on the Red List.

Econ. cost: (ag) denotes damage to agricultural crops and (fr) to forestry.

Groups: (A1) Species with a positive value, (A2) species with widespread positive value and limited negative value, (B) species with ambiguous values, (C1) species with large costs and small benefits, (C2) species with a negative value. Refer to the text for detail.

Notes:

a Paid by hunters to the prefectural government.

b Hunters killed approximately one thousand boars a year and the boar meat was sold at about 30 thousand yen.

The amount of damage to agriculture and forestry was based on subjective estimations claimed by owners of farms or forests.

presented live robins and jays to the shōgun (the hereditary generalissimo of the central government) in Edo (present-day Tokyo) as partial fulfilment of its financial obligations. The Amami woodcock *Scolopax mira* is another unique species, but generally attracts only bird watchers and photographers. Local people used to catch the Oriental white-eye *Zosterops palpebrosa* because of its attractive song and use it in an island-wide white-eye warbling competition. As current regulations are very strict, it is illegal to catch them. During the early 1980s, the local

Photo 2.3 Purple jay *Garrulus lidthi* (Photograph courtesy of H. Torikai).

government permitted about 500 captures per year, and at that time, I often saw the white-eyes singing in cages while walking along a street.

Another type of species in the group A1 is insectivorous birds. Nakayama (1980) estimated the benefits of forest birds to the red pine forests of Japan, where problems with pine beetles could become serious. He calculated their value in terms of reduced insect control costs and damage to timber at ¥108 billion (about US$720 million) per year for 2.27 million hectare of forests. If the forests in Amami were considered similarly, the value of insectivorous birds would amount to ¥3.32 billion per year, which is close to the total sales of the island's wood products in 1984 (¥3.90 billion). Bark feeders would make the greatest contribution and the Japanese pigmy woodpecker *Dendrocopos kizuki* is one of them. It is the smallest woodpecker in Japan, which is widespread and found even in large cities. Then follow other insectivorous birds with broader

Photo 2.4 Ryukyu robin *Erithacus komadori.*

feeding habits; the great tit *Parus major*, the varied tit *Parus varius*, the Narcissus flycatcher *Ficedula narcissina*, the black paradise flycatcher *Terpsiphone atrocaudata*, the ashy minivet *Pericrocotus divaricatus* and the Japanese bush warbler *Cettia diphone*. Nakayama (Ibid) emphasised that this was only part of the total value of forest birds because they also feed on agricultural pests and provide other benefits, such as aesthetic or recreational values.

The second group of species (A2) provides positive values to the local society in general, while imposing relatively small costs to agriculture and forestry. The Owston white-backed woodpecker *Dendrocopos leucotos owstoni* is the largest bark beetle predator and an endangered subspecies, while its pecking can damage shiitake mushroom logs. Mushroom cultivators put spores of shiitake on oak logs that also provide suitable habitat for insects. Feeding woodpeckers can damage the bark. However, covering the logs with fishnets can prevent such damage, and the government subsidised their cost. Zentoki Hajime, a local cultivator, told me that he did not submit any damage reports; the estimated amount of damage was small, only 0.45% of the total sales of shiitake mushroom in the island (¥76,716 thousand) in 1983. The other species in this group, Amami rabbit *P. furnessi*, has a very peculiar appearance for a rabbit with black or dark brown hair and other primitive morphological characters as described above. Individuals sometimes make a ticking sound apparently to communicate with each other; this is not very common among lagomorphs and scientists do not yet understand its function very well. The rabbit is recognised as a living fossil not only nationally, but also internationally. It was 1963 when the Japanese government designated it as a special natural monument, but it didn't establish any conservation measures for it other than prohibiting its hunting. It was considered as a pest in the 1980s among forestry managers as it ate saplings of sugi cedars. A village mayor said, "the rabbits bring one hundred harms but no benefits." Ironically, the cedars came to have no commercial value because of its low-quality wood. Cedars are no longer planted on the island, solving the problem without any economic costs.

2.2.2.2 *Species with ambiguous values*

Two species are difficult to determine whether their positive values are collectively larger than their negative values or vice versa. They are the Ryukyu wild boar *Sus scrofa riukiuanus* and Formosan green pigeon *Treron formosae*. Wild boars eat crops of tankan oranges, shaddocks, Japanese plums, sweet potatoes, sugarcane and peanuts. On the other hand, people hunt them for recreation using dogs and enjoy eating boar meat. Another benefit is tax revenue for the local government. The number of permits issued for trapping increased between 1985 and 2016, while that for gun hunting decreased during the same period, suggesting that the recreational value may have declined. In Kyushu and farther north, boar populations have increased so much that their damage to agriculture is regarded as a severe problem. People on Amami are more generous in their attitude towards the boar than those in Kyushu because they eat boar meat much more often; the local traditional food culture is different from that in Kyushu but shared with Okinawa. Hunters sell boar meat to butchers, who sell it for household consumption. There are many more such local sales channels in Amami than in Kyushu and Honshu.

The Formosan green pigeon is a subspecies endemic to the Nansei Archipelago and has another name, the shakuhachi pigeon. The shakuhachi is a Japanese traditional bamboo flute, which produces a serene sound somewhat similar to the song call of the pigeon. The same species is found in Taiwan, but here the bird has quite a different appearance, with bright red head feathers, while the subspecies in Japan has a dark green head. Japan's national government has protected this species from hunting. However, the bird eats Japanese plum fruit (*Prunus salicina*) and degrades its commercial value. Based on my own observations, the brown-eared bulbul, jungle crow and Oriental white-eye are also frugivorous and much more numerous. The pigeon does not quickly fly away, so farmers may observe fruit damaged by it more often than other birds. Farmers can estimate the monetary losses, apart from its accuracy, but its conservation value is intangible.

2.2.2.3 *Species whose negative values outweigh positive values*

The habu, the most dangerous species in the island, is a habitat generalist and can be found from residential areas to remote mountains. Its most favoured habitat is sugarcane fields, where black rats are most frequently found and eaten by habu (S. Hattori, personal comm). A local public health centre pays ¥3000 for a habu corpse as part of efforts to reduce its numbers and human injury. A merchant may purchase a live habu at a much higher price, even exceeding ¥10,000 (US$90) if it exceeds 1 m (Z. Hajime, personal comm). Local people caught around 10,000–12,000 habu from 1993 to 2017. There was a tourist place that showed a fight between a habu and a mongoose and sold various kinds of goods, such as liquor, medicine, wallets and belts, processed from dead habu. Also, many people catch snakes in the evening to earn money after their daily work. Anecdotal evidence is great that the snake generates a significant amount of economic benefits, psychological worries and physical injuries for local people. However, fortunately, the number of injuries is decreasing and the snake has killed nobody for the past three decades. Although the local government has spent a tremendous amount of money (e.g., ¥54 million in 2016) on a long-term project to control its population, the effectiveness of the project is not clear. The number of snakes caught in recent years has increased, and this may be due to a growth in either the habu population or hunting pressure.

Most of the species in the last group (C2 in Table 2.3) have a wide-ranging distribution, without having any specific conservation values and economic benefits, while they inflict considerable damages to crops. The jungle crow *Corvus macrorhynchos* injures primarily Japanese plum fruit and cabbage according to unpublished government reports. The brown-eared bulbul *Hypsipetes amaurotis* harms cabbage, tankan oranges and shaddocks. The damage caused by these two species is far greater than that caused by the other two in the same group, the eastern turtle dove *Streptopelia orientalis* and the black rat *Rattus rattus*. The areas damaged by these two species are also much larger than that of the others. The black rat feeds on stems of sugarcane, but causes relatively small estimated losses. The rat is the most important prey of the habu (Mishima, 1966; Shinjo and Nishimura, 1984). The rats, living not only in farmland but in houses, attract the habu which occasionally attacks people. Therefore, it imposes more evident costs to society indirectly through the habu than to agriculture. The rat also causes sanitary problems, serving as a disease vector. The turtle dove is widely distributed across the country, and farmers reported damage to a variety of crops, such as peas, corns, wheat and rice in Honshu (Terauchi *et al.*, 1985; Yamaguchi and Yoshida, 2006).

2.3 Impacts of logging and species survival from the 1980s to the early 1990s

2.3.1 *Forestry*

Island species populations are the most vulnerable to extinction. There have been many such extinction episodes around the world (Williamson, 1989). It is also true in Japan; all the extinct mammal and bird species endemic to Japan were islanders (Table 1.1). Thus, Stattersfield *et al.* (1998) nominated the Nansei Archipelago as one of the most urgent priorities among the 76 areas in the world for bird conservation. WWF International also ranks the Nansei Archipelago as a "critical or endangered" terrestrial ecoregion. Yet, less than 1% of the forests on Amami were protected from such development activities as logging, large entertainment parks and vehicle roads, before the government established a national park in 2017 with 11,541 hectare of forests—19% of Amami's forested area. The national government has also been endeavouring to have Amami's forests designated as a UNESCO World Natural Heritage site for some time, as described in the following chapter.

In the distant past, two types of forest managements were practised on Amami. One was coppice management conducted near residential settlements (satoyama), using wood to produce charcoal for fuel. The other was practised in remote mountainous areas (okuyama), where large trees were selectively cut for building houses or making dugout canoes. The demand for hardwood for industrial use increased after Japan began its military invasions and economic activities in China and Korea. The central mountains of Amami had no forest roads, so oxen dragged logged trees from the mountain slopes down to villages (T. Nakamura, personal comm). Broadleaf trees on the island such as *Castanopsis sieboldii* (Fagaceae: beech family) and *Schima wallichii* (Theaceae: tea family) were suitable not only for building houses but also for railway sleepers. The demand for hardwood increased during the period of military control over Korea and China (1904–1945). Then a company in Kagoshima purchased large areas of forests in the central mountains, equivalent to about 10% of the forests on the island. They started selectively cutting trees with a DBH of around 50–60 cm, to be used on the continent for railway construction (H. Tabuchi, personal comm). The amount of wood production for saw timber was relatively small in 1914, about one-fifth of that logged for saw timber and pulpwood combined in 1984. Loggers had not yet started clear-cutting for chip wood, and the cutting in the 1910s would not be expected to have had a severe impact on the survival of any species. The company never sold the forest land it purchased after making a large profit from its sales to the military. In the 1980s, the company's forests played a crucial role in species conservation as discussed below.

The United States occupied the Amami Islands after the Pacific War from 1945 to 1953. The forest age distribution in 1985 suggests that forests in the central mountains were left almost untouched during the period of occupation. It seems that people cut trees only in nearby woods. After the United States returned the islands to Japan in 1953, the national government provided substantial financial aid to the forestry sector, equivalent to 22% of the forestry sector's net production during the period from 1954 to 1963. The subsidy covered most of the cost for forest road construction, so commercial timber harvesting became feasible in the island's interior and wood production increased precipitously (Figure 2.2).

The Korean War stimulated demand for pulpwood from 1955, and sawlog production increased in the early 1960s (Fujita, 1984). The island's wood production peaked in 1963 and again in 1972, but declined suddenly within a few years after the second peak (Figure 2.2). The decrease was particularly large for sawlogs, suggesting a rapid decline in the area of old forests with large trees. The proportion of pulpwood in total wood production became outstanding. This occurred at the same time as clear-cutting of okuyama natural forests and expansion of coniferous plantations north of Amami, as described in Chapters 1 and 4.

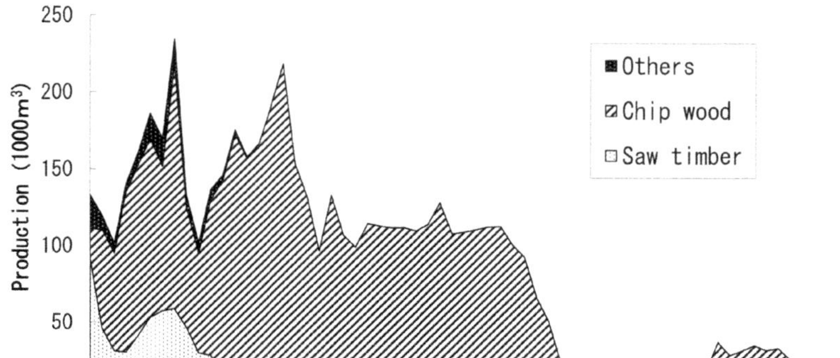

Figure 2.2 Changes in wood production on Amami Oshima.

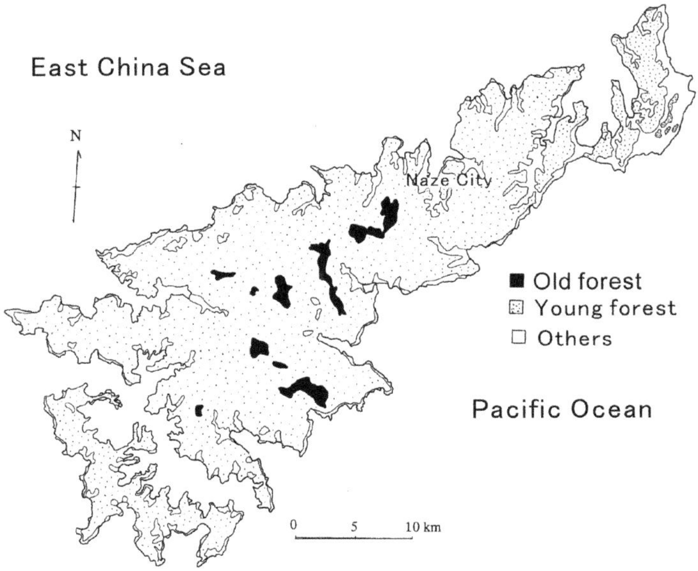

Figure 2.3 Forests that had not been clear-cut after 1954 (painted black) and young secondary forests (black dots) on Amami Oshima as at 1985.

Identifying the location of the old-growth forests from aerial photographs, ground observation and forest inventory records, I estimated the total area of unlogged forests and selectively cut forests aged 60 years or older. They occupied approximately 1800 hectare (2.6% of the forest area) in 1985—mainly small and fragmented remnants as a result of clear-cutting after 1954 (Figure 2.3).

I classified forest land owners into five types. Small individual owners are the largest in number as well as in the total area they own. Still, they are generally not interested in forest management because of small land size and because their occupations are not connected to forestry. The government and loggers often find it challenging to identify the owner of a forest stand. Yet loggers were able to find and cut trees down in the 1980s and earlier. The second and the third types, local settlement communities and municipal governments, sold standing trees to raise money for public purposes, such as reconstructing a public hall or renovating school buildings. These three groups received specific budgets for forest road construction from the national and prefectural governments and cut most of their forests earlier than the other two groups. They planned to manage trees on short rotation cycles of 35–40 years.

The fourth ownership type, the national forests, comprises 6% of Amami's forest land. The national Forestry Agency promoted clear-cutting until the early 1990s as part of its policy to enhance wood production. The Amami Forest Regional Office was under the jurisdiction of Kyushu Forestry Bureau.

As illustrated in Chapter 6, forestry managers in northwestern Kyushu were and still are enthusiastic about wood production while paying little attention to biodiversity conservation. On the other hand, officers in the local Amami office were not able to ignore the role of forests in flood control because the island had falls of rain of over 360 mm within 24 hours four times during the past 20 years, the heaviest of which was 647 mm (Kishō-chō, 2019). They decided to not cut strips of trees about 10-m wide on ridges and along streams, which resulted in maintaining about 30% of the forest cover. However, also they did not receive specific government subsidies for road construction, which the three groups above had. Thus, they had to follow the instructions from the Kyushu Forestry Bureau and continue clear-cutting only to accumulate a financial deficit. The turning point came in the late 1990s when a new director changed the primary management goal from maximising wood production to conserving the forests' unique species. There was no need to establish a new forest reserve or to change forest laws and regulations. He managed this by getting funds directly from the forestry agency office in Tokyo, which recognised the importance of biodiversity conservation values of Amami's forests for the first time. This new budget also partially supported the life of some people who were willing to work for wildlife conservation in Amami.

The last landowner group comprises the private company mentioned above, Iwasaki Sangyo Corp., whose head office is in Kagoshima City. While and after making a considerable profit before the Pacific War, it purchased about one-sixth of the island's forest land, coinciding with the core habitat areas of the unique species. It strived to create profits in two different ways from the other four groupings.

First, it started large-scale clear-cutting later than the others. In the middle of 1980s, either the Amami National Forest or Iwasaki managed most of the old-growth forests left on the island. By the early 1990s, the national forest agency ended logging except for three fragmented reserves totalling 390 hectare, so Iwasaki ended up having most of the remnant old-growth forests. Second, the company built a factory to produce floorboards for large buildings, such as gymnasiums, using large-diameter trees. Sustainable production of such wood requires forests managed on a long rotation cycle, which would leave large stands of old-growth forests. This contrasted with the other types of forest owners, who intended to manage trees on short rotation cycles. Therefore, the company was likely to control the fate of wildlife species dependent on large-diameter trees. However, they closed the sawmill in the 2010s, as described in the following chapter.

Large areas of young secondary forests were the consequence of whole-tree harvesting, which completely cleared forests, leaving no standing trees or fallen logs on a logging site (Photo 2.5). In contrast to the resulting young uniform secondary forests, relatively more valuable non-wood products, such as medicinal plants and precious ornamental plants, are found more commonly in old forests (H. Ōno, personal comm). Excessive clear-cutting was supposed to be prevented by the Forestry Agency's system of conservation forests. In this

Photo 2.5 Clear-cutting next to an old-growth forest.

system, the prefecture forestry department determines the areas of conservation forests in consultation and agreement with the landowners. One year's felling area is limited to 10 hectare in a catchment if it is in a conservation forest. The main problem with this system is that a catchment area is determined arbitrarily and is usually small, 100–200 hectare, so that loggers could cut out the whole area within 10–20 years. There was only one sizable area on Amami that a local municipality reserved for watershed management. Naze City (present-day Amami City) has such an area and decided not to cut previously selectively cut old-growth forests above a water reservoir. There are many other small community forests that the residents below have not allowed loggers to clear-cut.

2.3.2 *Unprofitable forestry and dependence on subsidies*

Certainly, profit-making was the strongest incentive for landowners, loggers and chip mill managers to log Amami's forests. All logging companies and chip mills were small, generally with less than 15 people and their financial situation was not very good (S. Shinmura, personal comm). The decline in the number and proportion of large trees and the consequent reduction in the potentially commercial biomass per area gradually made the forestry industry less profitable. Of the total value of primary products in 1985, pulpwood sales comprised 76.2%, and sawlogs, fire-wood, mushrooms and sharinbai (from which the dye for pongee is extracted) made up 4.4, 1.3, 4.9 and 6.6%, respectively (Sugimura, 1988). The importance of the pongee fabric as market wares is mentioned above.

The price of pulpwood, inversely proportional to its large share, was lower than that for sawlogs: ¥11,850 per m^3 compared with ¥14,000 per m^3 for the latter. The price difference was even larger for their secondary products: chips (¥16,800 per m^3) and sawn timber (¥58,864 per m^3) (Ibid). The production of sawn timber in 1980 fell to less than 5% of that in the peak year of 1963 (cf. Figure 2.2).

Most of the island's forestry products were shipped to the main islands of Japan (Honshu and Kyushu). The transportation cost from such a remote island to the main islands was even higher than that from mountain villages on the main islands, so it was difficult for the forestry products of relatively small commercial value to stay competitive with those produced on the main islands. Most of the trees ready to be cut had a low commercial value, as the young broadleaf trees and pines, which were overwhelmingly dominant, could be processed only for chips. Yet, the local governments advised forest managers to cut their trees on a 35–40 year cycle. They believed the largest amount (mass) of wood could be obtained with this length of rotation, based on estimates from the growth curve for each tree species. They did not consider the economic value of wood by volume at all. Nor was it a matter of interest to those forest managers whether or not the site quality or soil nutrient levels could be maintained using clear-cutting on such a short rotation.

Forestry in Amami yielded a "negative profit" as a whole when the financial aid from the government was taken into account. The national and prefectural governments provided about 80% of the financial support and the local government covered the rest in the 1980s (Sugimura, 1988). Using the year 1982 as an example, the net profit from forestry as reported in the government statistics was ¥1.27×10^9 in total, but this included the output from primary and secondary products. On the other hand, the financial assistance from the national, prefectural and local governments amounted to a total of ¥1.47×10^9 for the same year (unpublished statistics gathered by Kagoshima-ken Oshima-shicho). Without this assistance, forestry's net loss would have been ¥2.0×10^9 (Ibid). Most of the financial support was spent on the production of primary products, and forest road construction was the most extreme item. The national, prefectural and local municipal governments shared all the costs except for those covered by the Iwasaki Corp. and the national Forestry Agency.

The island's output of the primary products totalled ¥1.5×10^9 for that fiscal year, of which about 70% (¥1.05×10^9) was estimated to be the cost of production (Ibid). Therefore, it seems that about ¥2.5×10^9 (the financial support plus the costs) was spent to yield ¥1.5×10^9 of primary products. Also for 1982, raw wood production yielded a value of ¥1.17×10^9 while the cost of forest road construction was ¥1.12×10^9 so that the net profit of logging was far smaller than that expenditure.

The balance between forestry's production value and the amount of government subsidies was even worse during the early 2000s. Using the year 2008 as an example, the raw wood was valued at ¥53×10^6 and the expenditure on forest roads was ¥1023×10^6. The only way in which logging could be made economically feasible was through government subsidy.

Sugimura (1988) indicated that the dependency on government subsidies had become larger and larger as the growing stock and total amount of wood production decreased. Most of the money (i.e., 79.5% in 1983) was granted for the construction of logging roads, reflecting a great demand for public engineering works as a source of employment in the local society. The jobs provided by construction companies became more and more important relative to those created in forestry because of the subsidies. The ratio of the output from forest products to government subsidies became smaller towards the year 2000, before the government made amendments to adjust to the sharp decline in wood production (Figure 2.4).

Extensive clear-cutting not only causes soil erosion but produces a higher risk of flooding after heavy rains also, which once washed away a bridge on one of the main island roads. Consequently, engineering works for erosion prevention and flood control became essential. A large portion of the budget (totalling nearly 20% of the subsidy for forestry) has been spent on such construction works under the name of forest protection projects. This budget comes from a different category of government subsidy to that of the subsidy for forestry. The budget for this kind of public engineering works increased about tenfold every decade from the mid-1950s to the 1980s. During the same period, the government subsidy for forestry increased about threefold during the first decade and by a factor of 2.5 during the second decade. Then followed a continuous increase towards the peak—a value 61 times larger in 1998 than in 1954. In contrast, the output of primary forest products approximately doubled between the first and the second decade, 1955–1964 and 1965–1974. Then it increased only by 20% before dropping precipitously (Figure 2.4).

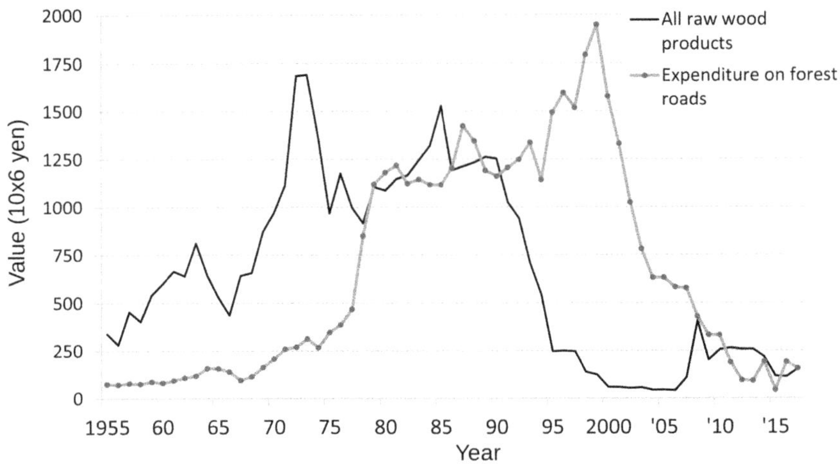

Figure 2.4 Annual trend in the production value of primary wood products and costs for forest road construction.

Most forest roads were abandoned after clear-cutting until they were repaired ready for the next logging operation. Logging on a short rotation requires a greater number of roads to be rebuilt, resulting in a greater demand for engineering projects. Short rotation forestry also provides more consistent logging jobs than long rotation forestry. However, it may keep wages a little lower than the longer cycle because pulpwood has less commercial values than sawlogs. Such a forestry economy had two income sources, one through the sale of forest products and the other from an external budget—the subsidies. I often heard locals criticising conservationists, arguing that Amami's unique wildlife species had generated little money and created no jobs. The reliance on public engineering works was not exceptional to Amami. Most of the Japan's local governments in rural areas adopted such strategies to secure job opportunities and to prevent young people from moving to large cities.

2.3.3 *Effects of forest cutting on wildlife*

Forests cover 85% of Amami, with this percentage remaining nearly constant since the first forest inventory published in 1954. They have provided vital habitat for most of the island's wildlife species. The area of forests on Amami prior to the end of the Pacific War was indeed smaller than that in recent years because terraced fields used to cover lower slopes above a village. Forests replaced the fields as island residents became more dependent on imported food, e.g., fresh vegetables from the main islands of Japan. Also, the population was larger before the war, when they were primarily dependent on local food production. Photographs taken by the US Army during the war showed terraced fields covering most of the slopes above the settled areas.

Increased exploitation of forest resources must have brought about profound changes in habitat conditions, particularly during the period from the 1960s to the early 1990s (Figure 2.2). Clear-cutting initiates secondary succession, so the restoration of wildlife habitats would depend on the rate and kind of forest regrowth and the habitat requirements of wildlife species. Therefore, it is essential to determine how long the restoration process takes to reach various forest conditions. In the 1980s and 1990s, Japanese pampas grass *Miscanthus sinensis* predominantly covered many clear-felled sites within a few years after cutting (Photo 2.6). This grass does not now cover as large an area as before; a greater variety of plant species are currently dominant. Broadleaf evergreen trees gradually replace the tall grasses and shrubs. Sugi cedar *Cryptomeria japonica* used to be planted but its growth or quality of wood was not good, so that all logged sites are now naturally regenerated. The evergreen oak, *Castanopsis sieboldii*, used to be very abundant, as it germinated more prolifically than other tree species. These days the camphor tree, *Litsea cubeba*, dominates for several years after clear-cutting. The height of the stump of trees cut by loggers is now lower than previously because wood prices are lower, and loggers need to take as much wood as possible to make the logging economically feasible. I suspect that logged *Castanopsis* trees do not have sufficient stump height for regeneration.

Photo 2.6 Clear-cut slopes covered with Japanese pampas grass *Miscanthus sinensis*.

Since *Litsea cubeba* is a pioneer species, current logging may produce a significant delay in the successional process relative to the past. In the late 1980s and early 1990s, when forests were frequently clear-cut, there was a wide range of successional stages of forests, from clear-felled areas to uncut forests. Then, it was possible to monitor and compare wildlife species populations across the variety of seral stages to examine the effects of clear-cutting-based forestry.

I grouped the forests in my study areas into four successional stages, primarily according to the average DBH of twenty to fifty canopy and subcanopy *Castanopsis sieboldii* trees. The stand age and average DBH used to define the groups were as follows: clear-felled sites (less than 8 years after clear-cutting or domination by *Miscanthus sinensis*, average DBH of less than 2.0 cm); young secondary forests (10–25 years after clear-cutting, average DBH of 5.0–10.0 cm; old secondary forests (40–63 years after selective-cutting or 70 years after clear-cutting, average DBH of 14.0–26.0 cm; primary forests (uncut or more than 59 years after selective-cutting, average DBH of over 30.0 cm).

2.3.3.1 *Birds*

Birds often serve as a useful biodiversity indicator, especially during the breeding season. Since birds sing most frequently after sunrise, we would begin our census at sunrise and visit as many points as possible by 11.00 AM. I worked alone counting birds before being supported by the Amami Ornithologists Club from 1993. We repeated the survey every eight years until 2018. We always attempted to use the same locations to survey populations in the breeding and winter seasons, though some of these places are no longer accessible. Bird numbers were estimated by the sample-counting method (Bond, 1957). We established sampling stations along logging roads and stayed at each for 10 minutes. The first observation point was about 500 m from the edge of a selected forest stand, as measured along the local road. Because the landscapes have numerous small valleys and ridges, no roads were straight. Then each succeeding point was located 500 m ahead of the last. The number of sampling stations selected in clear-felled areas, young secondary forests, old secondary forests and primary forests, were 5, 13, 15 and 14, respectively, in the summer of 1985, and 9, 12, 15 and 17, respectively, in the winter of 1985–1986. Each station was visited four to five times for 10-minute stays. At each station, we kept track of each bird as it moved around and attempted to count it only once during each stay. Birds flying high above the site, only to pass over the area, were not counted. Counts were not made on rainy days or on days when strong winds reduced bird activity.

The surveys showed that the following five species represent the different impacts of the clear-cutting regime. They are the Ryukyu robin *Erithacus komadori*, the Owston white-backed woodpecker *Dendrocopos leucotos owstoni*, the Amami thrush *Zoothera dauma amami*, the purple jay *G. lidthi* and the brown-eared bulbul *Hypsipetes amaurotis*.

The thrush, woodpecker and robin exhibited substantial population declines as the forest age decreased. The 1985 and 1986 surveys recorded the thrush only at three old forest plots, two of which had not been cut while the third was a 63-years-old selectively cut (but never clear-cut) stand. There were only seven individuals observed during the seven days census in the spring of 1986; one was counted during the summer of 1985 and another in the following winter.

Similarly, although to a lesser degree, white-backed woodpeckers were observed almost exclusively in the old forests (Figure 2.5). They were found only in young regrowth stands that contained old dead snags. It was much easier to observe a woodpecker in summer than in winter because males tap tree trunks in the breeding season. The Ryukyu robin provides a good example of negative

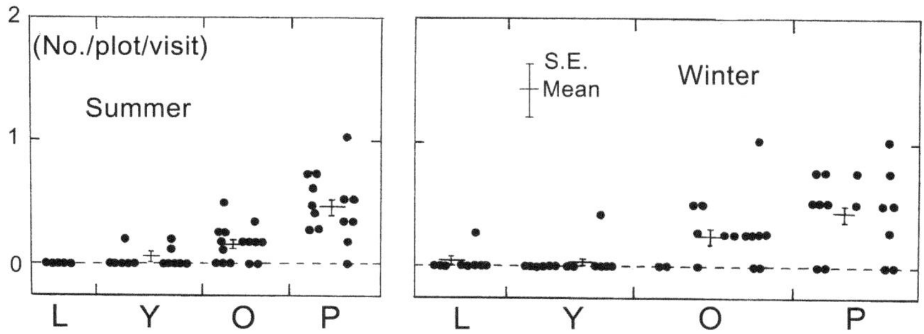

Figure 2.5 Average number of white-backed woodpecker *Dendrocopos leucotos* counted per census plot per visit (L: clear-felled site, Y: young secondary forest, O: old secondary forest, P: primary forest; Refer to the text for the explanation of these forests).

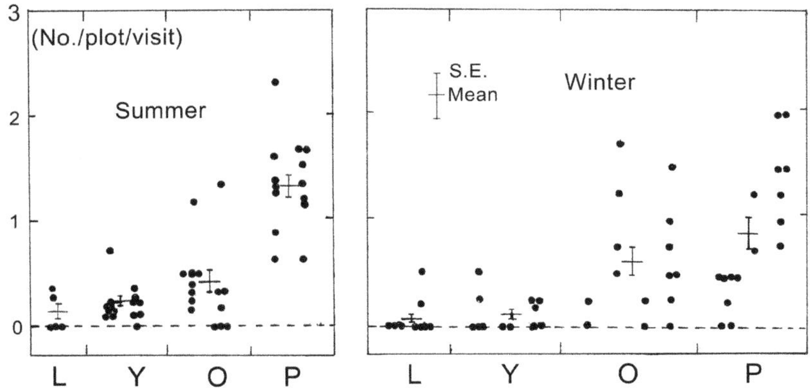

Figure 2.6 Average number of Ryukyu robin *Erithacus komadori* counted per census plot per visit (Abbreviations for L, Y, O and P are identical with those in Figure 2.5).

forestry impacts. Clear-felled sites and secondary forest did not provide suitable habitat either in the breeding or winter seasons (Figure 2.6). Robins sometimes flew back and forth between an old forest and a logged site or a young forest. Also, I did not find them in old forest patches isolated and not connected to others.

In contrast to these three species, the purple jay was as common in young secondary forests as in older forests both during summer and winter, although it, too, was rare in clear-felled sites (Figure 2.7). Because of the restrictions imposed by the government on the size of cuts in conservation forests, clear-felled sites were not very large, and clear-cuts grew to young secondary forests with a closed canopy in about 10, or a little over 10, years. Most of the forests in the central mountains had been protected from extensive clear-cutting for

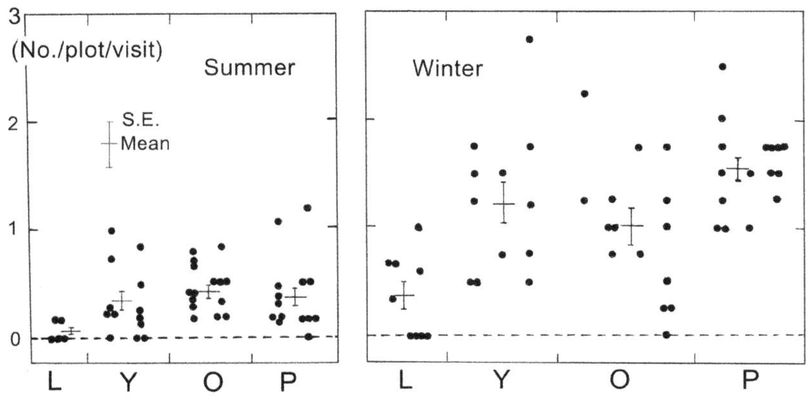

Figure 2.7 Average number of purple jay *Garrulus lidthi* counted per census plot per visit.

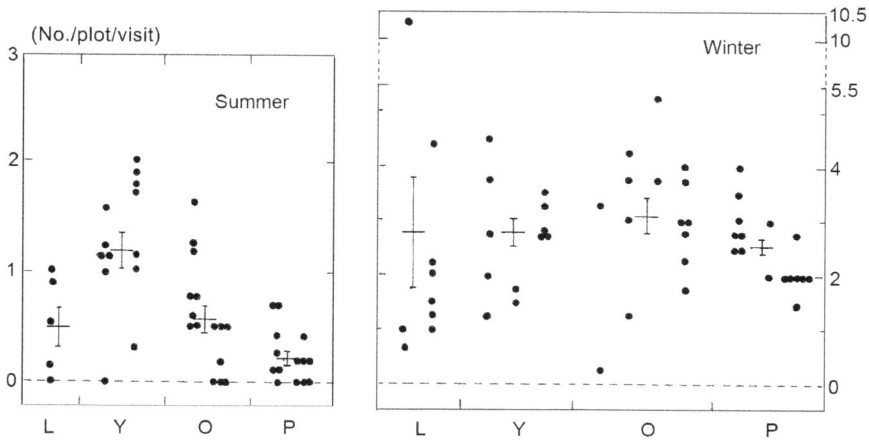

Figure 2.8 Average number of brown-eared bulbul *Hypsipetes amaurotis* counted per census plot per visit.

watershed protection, except for the Iwasaki forests. This was because the company did not agree with having such conservation measures, when it planned to cut trees. The fortunate thing was that Iwasaki Corp. was situated on the East China Sea side of the island and it had not started extensive clear-cutting on the Pacific coast side where there were larger areas of old-growth forests. Hideki Tabuchi, who was in charge of its forests, told me that the former president of the company was conscious of preserving forests for the sake of local communities downstream from its forests.

The direction of the effect was opposite for the bulbul in such a way that it was observed most frequently in young secondary forests in summer (Figure 2.8). In late autumn, the migratory bulbul arrived in large numbers from the north, and produced substantial damage to local orchards. Many frequented old forests, but population levels in the forests fell in spring.

Since selectively cut forests and uncut forests contained many large trees (Photo 2.7), I categorised them as old forests so I could compare them collectively with young secondary forests and clear-felled sites. Until recently, the old forests had distinctive differences in DBH, height and forest structures compared to younger stands that were clear-cut after 1953. I compared the average number of each bird species observed per census plot in old and young stands to estimate the magnitude of the population change for each avian species caused by the conversion of old forests to younger stands. The average number of individuals per plot in the old forest stands divided by that in the young stands gave the magnitude of the effect for each species. The forestry plan was to cut the forests over about 40 years, and the forest canopy would be restored in about 10 years. Thus, I assumed that young secondary forests accounted for three-quarters of existing young stands. Since the area of primary forests and that of old secondary forests were about equal, the average of these two stages

Photo 2.7 Selectively-cut forest with large trees such as *Castanopsis sieboldii* and *Schima wallichii*. The sign on the right says that this is a conservation forest for watershed management.

was taken as the population level in the old forests. The larger the value, the more the species was considered to be dependent on old-growth forests. This index is much simpler and easier to use to estimate the impacts of logging than the values presented in Figures 2.5–2.8.

Table 2.4 gives the results of the above process. The species at the top of the table were impacted adversely the most by clear-cutting. A total of 13 species have negative effects, while 6 species seem to have positive effects. What kinds of birds would be negatively affected, and vice versa? The species with great adverse impacts are cavity nesters; woodpeckers, robin and tits. This is to be expected because young trees are strong enough to prevent bacteria from decaying their trunks to form hollows. These bird species are also insectivores, except for the robin which feeds on ground-dwelling invertebrates. Other ground feeders—the thrush, woodcock and ouzel—prefer to live in old forests, but not the wood pigeon. The Narcissus flycatcher feeds in open spaces under a closed canopy (Sato *et al.*, 2011), as do other species of flycatcher. One thing important from a conservation point of view is that all the species endemic to the Nansei Archipelago seem to be impacted adversely by logging. Table 2.4 suggests that 12 of the species of insectivores in Table 2.3 other than the bush warbler would be adversely impacted. Ecologists have not clarified the benefits of living in mature forests for these species. Still, probably the key factors are the availability of decaying or old trees that the woodpeckers feed on and open space that the other species like to use as well as insect abundance.

Table 2.4 Assessment of the effects of conversion of old forests to younger stands on avian species according to the difference between the number of individuals observed in young stands and old growth

Common name (Scientific name)[a]	*Season*	*Population level*		*Population change*
		Old forest	*Young stand*	
Amami thrush	Spring[b]	0.36	0.00	0.00
Narcissus flycatcher	Summer	0.365	0.023	0.063
White-backed	Summer	0.31	0.028	0.090
woodpecker	Winter	0.35	0.030	0.086
Japanese pigmy	Summer	0.69	0.028	0.087
woodpecker	Winter	0.32	0.028	0.087
Amami woodcock	Summer	0.23	0.038	0.17
Formosan green pigeon	Summer	0.20	0.034	0.17
	Winter	0.13	0.030	0.23
Ryukyu robin	Summer	0.86	0.20	0.23
	Winter	0.75	0.11	0.15
Black paradise flycatcher	Summer	0.15	0.028	0.19
Ashy minivet	Summer	0.19	.059	0.31
Ruddy kingfisher	Summer	0.22	.076	0.35
Pale ouzel (*Turdus pallidus*)	Winter	1.12	0.53	0.47
Great and varied tit	Summer	3.01	1.58	0.52
	Winter	1.20	0.69	0.57
Purple jay	Summer	0.40	0.27	0.67
	Winter	1.27	1.01	0.80
Eastern turtle-dove	Summer	0.18	0.10	0.56
	Winter	0.39	0.42	1.1
Jungle crow	Summer	0.34	0.49	1.4
	Winter	0.48	0.46	0.96
Oriental white-eye	Summer	0.38	0.55	1.4
	Winter	1.19	1.33	1.1
Japanese bush warbler	Winter	1.54	1.82	1.2
Buntings (*Emberiza* spp.)	Winter	0.32	0.55	1.7
Black wood pigeon (*Columba janthina*)	Winter	0.60	1.20	2.0
Brown-eared bulbul	Summer	0.41	1.04	2.5
	Winter	2.77	2.71	0.98

Notes:

a Scientific name is given after the common name if not provided in the text or the other tables.
b I conducted supplementary census considering the lack of information.

2.3.3.2 *Mammals*

Of most concern among the mammals are the long-haired rat *D. legata*, the Amami spiny rat *Tokudaia osimensis* and the Amami rabbit *P. furnessi*. We seldom saw the long-haired rat and our trapping success rate was very low in the 1980s and 1990s, but Toru Nakamura, a forestry manager who retired in the middle of 1980s, told me that he often saw small groups of them moving from one tree to another. That was a time when there were plenty of large old trees.

I collected information on the location of forests where the rats had been observed. Of eight sightings, four were in uncut forests, and the others were in old forests selectively cut around 50–70 years before the observation. Of these, four were alive, three on trees and one on the ground, and the other four were found dead. Ikehara *et al.* (1981) indicated that large trees are necessary for the rat's nest-sites and they depend on acorns for food, which are scarce after large-scale logging operations. Figure 2.9 illustrates that forests younger than 20 years provide a smaller number of acorns compared to older forests in a good acorn year. These observations suggest that clear-cutting of old stands, possibly older than 50 years, would have great adverse impacts on the survival of the long-haired rat. Figure 2.3 also indicates that older forest that could provide acorns were small and fragmented.

In 1985 and 1986, I selected 23 (in summer) and 22 (in winter) logging roads to compare the number of pellets of the Amami rabbit between the four successional forest stages. The rabbits left a varying number of pellets in piles in open spaces such as logging roads or riversides. Pellets are often not found on the forest floor. The piles of pellet on logging roads were the most useful sign to indicate the abundance of rabbits in the field. I visited each of the chosen logging roads every two or three days and cleared all the pellets on the roads. For a total of three or four times, I counted the numbers of individual pellet on the roads. The field study was conducted in summer (from June to August 1985) and in winter (from November 1985 to February 1986). The pellets were found most frequently and consistently along roads in young second-growth stands (Table 2.5) with statistically significant differences to roads in the older forests (at the 1% level in the summer and the 2% level in the winter, using the two-tailed *t*-test). The rabbits also quite frequently used clear-felled sites, where grasses were abundant, during the summer. The results suggest that replacing old forests with younger stands would increase the abundance of the rabbit.

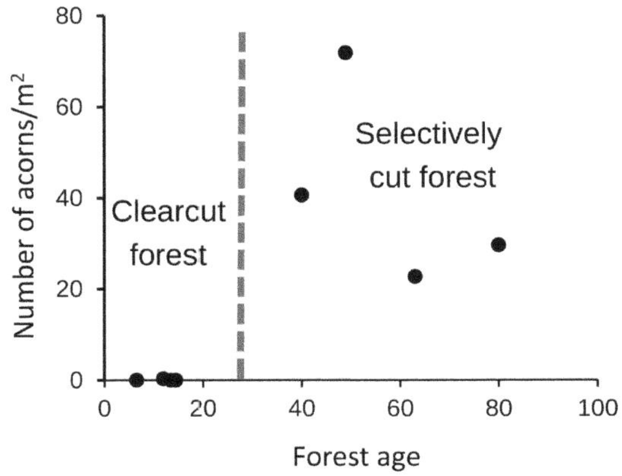

Figure 2.9 Acorn abundance per square meter on the forest floor after the years of clear-cutting.

Table 2.5 Number of Amami rabbit pellets found in the forests of various ages along road survey routes

	Stand age (years)	*Distance (km)*	*Pellets (no./km/day)*
(Summer 1985)			
Clearfelled site (Sample size: 2)			
Range	3–8	1.6–2.7	0–141
Average	5.3	2.2	70.5
SD	2.1	0.8	99.7
Young secondary forest (13)			
Range	10–21	0.8–4.8	0–276
Average	14.1	2.2	87.6
SD	3.1	1.3	76.9
Old secondary forest (4)			
Range	40–70	1.0–3.2	0–6.3
Average	54.7	2.5	3.1
SD	14.6	1.0	3.6
Primary forest (4)			
Range	59–Uncut	0.8–3.9	16–34
Average		2.6	23.5
SD		1.3	7.9
(Winter 1985–1986)			
Clearfelled site (3)			
Range	1–15	1.3–2.4	
Average	8.0	1.9	0
SD	5.5	0.6	0
Young secondary forest (13)			
Range	10–21	0.9–3.8	0–149
Average	14.3	2.2	48.6
SD	2.9	1.2	42.8
Old secondary forest (3)			
Range	40–70	1.4–3.2	0–6.3
Average	54.5	2.2	5.9
SD	14.3	0.9	5.3
Primary forest (3)			
Range	59–Uncut	0.8–5.3	16–34
Average		3.3	15.3
SD		2.3	26.6

Abbreviation: SD, standard deviation.

Source: Sugimura (1988).

The young secondary forests had richer undergrowth cover than the older forests and the roadsides in the younger forests were covered more densely by grasses and ferns, where many signs of feeding were found.

However, the results also revealed the importance of mature forests. The distribution pattern of the pellets along roads in the young secondary forests suggested that areas near the borders between old forests and young secondary forests support substantially higher population densities of the rabbit in winter than areas situated relatively far from the old forests. This was further confirmed by the larger number of pellets along logging roads in young secondary forests in the vicinity of adjoining mature forest (primary forest or old secondary forest) than in areas far from mature forests (Table 2.6). In 10 out of the 11 sites compared, there were more pellets near the mature forests in winter. This comparison indicates that the rabbits favoured areas near boundaries between young second-growth stands and mature forests as a winter habitat (significant at the 5% level using the two-tailed t-test). Pellet counts in summer did not produce a similar result (Table 2.6); the availability of acorns may have caused such a seasonal difference.

Thus, the logging conducted to that time might have favoured rabbit populations in the central mountains, where primary forests and old secondary forests were interspersed with young secondary forests. The continuing decline in the area of mature forests could quite possibly cause a further reduction in Amami rabbit populations. However, predation by the invasive mongoose complicated an analysis of the further decline in rabbit populations, as it was difficult to analyse these two factors separately.

The impact of logging on spiny rats was also ambiguous. I trapped some live spiny rats in the late summer of 1985 and winter of 1986 on transect lines through forest stands of various ages. I used 12–34 trap stations in each stand at 20-m intervals. A total of 44 spiny rats were caught, of which only one was in a clear-felled site. There was no significant correlation between the forest age and the number caught in the young or old forests. The results suggest that the

Table 2.6 Comparison of the frequency of faecal pellets (number/km/day) in young forests between (a) near an old forest and (b) more distant from an old forest

Pairs (Winter)		*Pairs (Summer)*	
Nearby	*More distant*	*Nearby*	*More distant*
–	–	153	17
10	0	107	90
7.3	0	0	90
10	7.3	107	0
107	99	70	43
187	119	297	86
177	0	–	–
52	13	79	41
77	13	35	41
41	13	–	–
22	10	48	30
61	74	24	65

"Pairs" means that they are on the same roads in a young forest compared to a nearby old forest and a more distant old forest in summer and winter.

rat would be negatively affected by logging for a time before the area is covered by young secondary forests when its population would return to, or may even exceed, its original level. Mitsui and Ikehara (1979) trapped the same species in evergreen oak forests but not in agricultural fields on Okinawa Island, which also indicates that this species is dependent on a forest habitat. The rat is dependent on acorns for food, so the reason why it was more numerous in young forests than in old forests is not clear.

2.4 Towards sound human-wildlife interactions

Self-supporting industries make up only a small portion of Amami's economy, while government subsidies create a great number of jobs—particularly through public works. Forestry was and still is subsidised by the national and prefectural governments in no small degree. The cost of building logging roads has been free for most forestry operators. As a result, young secondary forest and clear-felled sites replaced most of the mature forests by the early 1990s. The government planned to manage those young forests with a 35–40 years' rotation cycle of logging for pulpwood, even though the latter had relatively low commercial value.

The Amami thrush, Owston white-backed woodpecker and long-haired rat must have all suffered substantial population declines because of the extensive clear-cutting. Fortunately, a change in the wood economy in the early 1990s saved these species, as described in the following chapter. For another three species, the Amami rabbit, purple jay and spiny rat, the actual impacts of clear-cutting are less obvious. Yet, a further reduction of the old forests seems to have caused a reduction in rabbit populations. The adverse effects of logging on the purple jay and spiny rat should not be severe, because the area of logged stands was far smaller than that of young secondary forests. Although their population levels in clear-felled sites were low, they were as numerous in young secondary forests as in older forests.

One way of avoiding the severe impacts of clear-cutting is increasing the length of forest rotation, i.e., lengthening the time between logging events. This would significantly reduce the probability of the local extinction of some arboreal species. At the same time, the subsidies for forestry could provide jobs for local people and thus maintain the contribution of the forestry industry to the local society. Harris (1984) proposed a similar model in which long rotations for logging would maintain the habitats of wildlife species dependent on old forests. The rotation time can be extended without reducing the present level of the financial output of forestry because the value of timber tends to increase with forest age. Construction of logging roads itself does not produce economic profits, so reducing the number of new roads could save substantial amounts of money through reduced subsidies. The saving could be invested in an investigation of more beneficial uses of mature forests. In this way, the amount of forest logged could be reduced without diminishing the current role of forestry in the local economy.

The most popular approach taken by biologists, on the other hand, has been the establishment of nature reserves, which prohibit any human activities that are likely to have adverse effects on the unique biodiversity (e.g., Diamond, 1975). They emphasise the merits of species preservation without paying much attention to benefits obtained from other human activities. However, such species preservation methods are not very practical in many cases, because of the large areas of reserves needed to satisfy many biologists.

The difficulty for conservation values in competing against economic priorities was well illustrated in case of Amami, where its three forest reserves totalled only 0.53% of the island's entire forest area in the 1980s and 1990s. The values of unique species were intangible and difficult to quantify, so it was hard for policy decision-makers to introduce forest protection measures by prioritising conservation over human activities whose benefits are visible and readily quantifiable. An unexpected change in circumstances occurred in the early 2000s when Japan's Ministry of the Environment decided to propose Amami as one of the candidate areas for World Heritage status.

Supplemental notes

2-1) Amami City was formed by the merger of three municipalities, Naze, Kasari and Sumiyo, in March 2006. Therefore, the city names in the text differ according to the timing of each event.

2-2) Amami Oshima has a county branch of Kagoshima-ken (prefecture), which issues "Amami Guntō no gaikyō" every year and this provides a large quantity of statistical information from 1955 to the present. The statistical figures cover the Japanese fiscal year, from 1 April to 31 March.

2-3) The previous average weather conditions were taken from a website of the Japan Meteorological Agency. Available at: http://www.data.jma.go.jp/obd/stats/etrn/index.php?prec_no=44&block_no=47662. (accessed 13.07.19).

2-4) Government in Japan publish annual statistics in terms of fiscal years that cover the period from 1 April to 31 March. There are no exceptions to this practice, so I omit "fiscal" whenever I provide annual statistics because all the statics in this book are from government sources.

2-5) Available at: http://kyushu.env.go.jp/okinawa/awcc/report.html. (accessed 15.07.19).

2-6) Available at: http://www.tbg.kahaku.go.jp. (accessed 15.07.19).

2-7) Available at: https://www.pref.kagoshima.jp/bc05/hakubutsukan/iimono/areanature/documents/44970_20150411175031-1.pdf. (accessed 15.07.19).

2-8) A European type of government was established in 1868 and before this period Amami was a territory of Satsuma, a local government. Satsuma pledged its fidelity to the Shogun, who ruled Japan at that time and imposed taxes on the lord of each local government area. The capital was located in Tokyo, which was called Edo at that time.

3 Amami Oshima (2)

Controlling mongoose populations and a new period of nature conservation

3.1 Another threat to wildlife: mongoose invasion

As Figures 2.2 and 2.4 illustrate, forestry practices became economically unprofitable in the 1990s and resulted in a precipitous decline in log production. The reason was straightforward. In the late 1980s, the Japanese yen surged against major currencies. One US dollar was equivalent to around 250 yen in the first half of 1985, while it went down to about 130 yen in 1991 (Nihon-ginkō, 2020). Penna (2002) explains that the chip millers' situation worsened throughout Japan and they were stuck in a cycle whereby the relatively higher costs of domestic pulpwood contributed to the paper companies' declining use of domestic products. Japan's Forestry Agency supports forestry managers financially for planting, weeding and thinning, as long as they are willing to develop commercial tree plantations. The Japanese government does not subsidise the income of producers through direct compensation. So, many forest owners were able to sustain plantation forestry in Kyushu, Shikoku and Honshu, but this was not true of Amami, and all the island's chip mills closed during the 1990s. However, civil engineering works did not suffer the same way, because road building continued outside forested areas. After all the mills closed, the area of old forests still declined to some extent because a small amount of wood was exported to Kyushu. I estimated the area of old forests at 15,700, 9800, 7300 and 6300 hectares in 1970, 1980, 1990 and 2000, respectively (Sugimura, 2002). Nevertheless, much of the forest area cut in the 1960s and 1970s can be expected to regrow to a condition similar to the remnant old forests. However, another serious problem found its way into the central mountains. This was mongoose invasion and their enormously increased numbers.

Before the mongoose invasion, conservationists recognised the need to control feral cats and dogs. A total of 11 of the 15 dog scats and some cat scats that I found on forest roads in areas with high rabbit numbers contained a lot of rabbit. Takeichirō Minami (personal comm), then a professional habu hunter, occasionally observed a feral cat killing a rabbit that was trying to escape by running up a slope. He suggested that the rabbits would be more vulnerable to cats than dogs since dogs would not be able to climb such steep slopes. When counting rabbit pellets along forest roads in 1985 and 1986, I seldom found them near residential areas. They were also scarce around the most populated area of Naze City (currently Amami City) and more numerous in the central and western part of the island, where the human population density was low relative to the northern part. The absence of the rabbit pellets near residential areas, particularly close to Naze, suggested the following relationship between the number of rabbit pellets on the road and the road location. The number of rabbit pellets was positively correlated with the distance from residential areas and negatively correlated with the population size of nearby settlements. The population size of feral cats and dogs is potentially large in areas where there are many people. I prepared a multiple linear regression analysis and was able to obtain a statistically significant regression model which supported this hypothesis.

It was 1979 when a group of Amami residents released 37 small Indian mongoose *Herpestes auropunctatus* on a hill adjacent to Akazaki (Red Cape) facing the East China Sea. This place is less than 1 km north of the residential area of Naze City. There was a cottage in this area where school children stayed overnight during summer for education and training outside the school. A local had many mongooses that he used as part of his show business activities, and he believed that the mongooses would be able to kill any habu snakes in the surrounding fields, thus helping to protect the children. The release was reported favourably in the local press.

The release point is about 5 km north-northeast of the northern end of my study areas, where I monitored wildlife populations for more than 30 years. On my daily visits to my study area during 1985 and 1986, I would pass through a small valley located about 5 km south of the spot where the mongooses were released. However, I only recall seeing any twice, and I misidentified them as weasels *Mustela itatsi*, because I knew that a famous Japanese biologist had released some to control black rats. It was 8 years later when I saw at least one or two individuals every time I went through the valley, which had been extensively clear-cut through the 10–15 year prior to 1985. This small valley suffers from strong northerly winds in winter. Even the regenerating young trees did not grow higher than pampas grass *Miscanthus sinensis* for another 10 years, but the mongoose survived. I selected this area as representative of the island's logged sites for my bird counts (Figures 2.5–2.8). My observations of the mongoose here conform with the distribution area that Ishii (2003) estimated based on interviews with local people, so it is certain that the mongoose distribution range was restricted to small areas at that time.

The mongoose suddenly expanded its distribution during the early 1990s. Its range covered an area approximately 10 km radius from the release site by 1989 and an area with a 20 km radius by 1997, so encompassing the mountainous areas containing many threatened species (Yamada and Sugimura, 2004). Kankyo-chō (1999) estimated the rate at which the mongoose extended its range at about one kilometre per year and the population to be 5000–10,000 in 1999. Watari (2008) reported food preference determined by digestive tract content analyses of dead mongooses. He identified 78 different items, species and many unidentified species, including endemic species such as the Amami rabbit, spiny rat *Tokudaia osimensis*, purple jay *Garrulus lidthi* and Ryukyu robin *Erithacus komadori*.

However, the government failed to take immediate action. Instead, it was local conservationists who anticipated that it might become a severe problem and began surveying mongoose distribution and abundance in 1990 (Abe *et al.*, 1991). This small group of amateur scientists used heavy traps and set them along forest roads, while the mongoose rapidly expanded its range deep into forests away from the roads. The roads run mostly along ridges so that they were only able to catch a small portion of the whole population that must have extended down to valley bottoms.

The government began a control project in 2000. The range covered more than half of the island in 2002 (Ishii, 2003; Figure 3.1). Fukasawa *et al.* (2013) estimated the number of mongooses to have peaked at about 10,000 around the year 2000, which was more than 200 times the number originally released. I thought it was too late to take any action because I saw them almost every day,

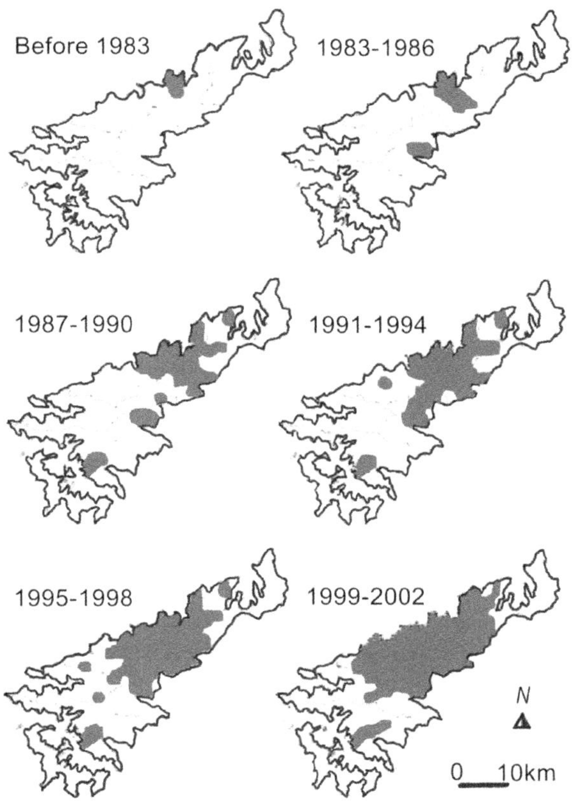

Figure 3.1 Expansion of mongoose distribution from the 1980s to the early 2000s (Adapted from Ishii (2003). Original images: Courtesy of the author.)

sometimes an adult with a couple of juveniles, even in areas far from the 1979 release point.

The government paid volunteer trappers a maximum of ¥2200 (ca. US$20) for each animal caught, which served as an excellent incentive to earn extra income, whether or not one had a strong desire to eradicate the mongoose. It worked as long as the mongoose was abundant and there were not many trap-shy individuals. The biggest challenge at the beginning was that the cage traps (live traps) were too heavy to carry up and down the slopes. It soon became apparent that cost-effective trapping would end sooner or later. Then T-shaped cylinder traps (kill traps) were developed. These were much smaller and lighter than cage traps so trappers were able to carry many more traps. The next step the government took was to form a team of trappers to be employed on a small salary. The Ministry of the Environment (MoE) has a small budget relative to the other ministries, so the amount of money available to pay for trapping was inevitably small. Then, the government decided to employ as many trappers as possible, instead of securing sufficient funds to pay for each trapper to live on the island. Unfortunately, living costs were relatively high because of the island's dependence on imported food and household items from other regions of the country. Yet many passionate young people came voluntarily to the island to eradicate the invasive mongoose despite such low levels of pay and short periods of employment. The team was called "the mongoose busters," and at one time had more than 40 members (Kankyō-shō, 2014).

During the 1990s and the early 2000s, working on the island and seeing so many mongooses, I was almost desperate. However, quite unexpectedly it did not take many years before the number decreased to such a low level that I seldom saw them in the 2010s. This was undoubtedly because of the efforts of "the busters." I saw these people in the mountains much more often than I saw employees of road and dam construction projects. Some of them even said that they were willing to sacrifice their life for the mongoose eradication. Fukasawa *et al.* (2013) estimated that the mongoose numbers had dropped to less than 200 in 2011.

The government brought in dogs and a handler from New Zealand during 2008 as the final stage in eradication. The dogs were trained to find mongoose faeces so traps could be set in the vicinity. I understand there were two dogs being used in 2014, one working in the areas where the ministry considered eradication of the mongoose had been successful. The other dog was supposed to work to find faeces. The first dog became bored with its work because it was not finding any signs of a mongoose. The solution was to change the positions of the two dogs from one another (M. Kimura, personal comm). The detection efficiency of the dogs was higher than the efficiency of kill traps (Mitani *et al.*, 2014). Table 3.1 illustrates the successful control of mongoose after 2014 to 2018, the most recently reported period. The number of captures decreased from the peak of 2702 in 2006 to only 1 in 2018. A comparison of the number of mongoose captured by traps and dog handlers in 2014 and 2015 in Table 3.1 indicates the excellent efficiency of the dog handlers. Because the small number

Table 3.1 Number of mongoose captured by traps and dog handlers on Amami Oshima

Year	*Catch number*	*Trap nights*	*CPU*	*Catch by dog handlers*
2003	567	76,667	7.40	–
2004	699	177,438	3.94	–
2005	1191	497,797	2.39	–
2006	2702	1,039,970	2.60	–
2007	782	1,378,069	0.57	–
2008	931	1,899,238	0.49	1
2009	565	2,023,762	0.28	0
2010	282	1,939,964	0.15	1
2011	245	1,872,878	0.13	11
2012	179	2,263,076	0.08	18
2013	110	2,482,528	0.04	20
2014	39	2,597,407	0.015	32
2015	22	2,724,129	0.008	18
2016	27	2,671,356	0.010	1
2017	10	2,760,923	0.004	0
2018	1	2,598,955	0.0004	0

Trap nights: number of traps effective × number of nights of trapping.

CPU: the number caught per 1000 trap nights.

Source: Kankyō-shō (2014), Okinawa Amami Shizen Kankyō-jimosho[(3-1)].

of released mongooses expanded to such a large number, we can expect that monitoring is required for the time being at least, but possibly forever.

3.2 The impact of the mongoose on the Amami rabbit

Since the island's mongoose population increased rapidly and they preyed on rabbits, they were very likely to have lowered the rabbit population. I conducted an extensive monitoring survey, mainly with students from Kagoshima University, from January 1993 to December 1994 (Sugimura *et al.*, 2000). We found signs of Amami rabbits in forests (mature and regrowth) and clear-felled sites dominated by perennial grasses. We included areas from all these vegetation types in our survey, except for small patches completely isolated by paved roads, agricultural fields and residential areas. In total, the survey took up 292 person-days, more in winter than in summer. The surveys on Tokuno Island were conducted in March 1995 and took 55 person-days. We searched for pellets while walking along 92 km of forest roads and 222 km of streams, as well as driving along 417 km of roads. We divided the whole of the rabbit's distribution into 14 regions, which were separated by highways, rivers or major forest roads (Figure 3.2). The survey routes covered the entire distribution range. We used pedometers to measure distance when we searched on foot. Because the habu is generally

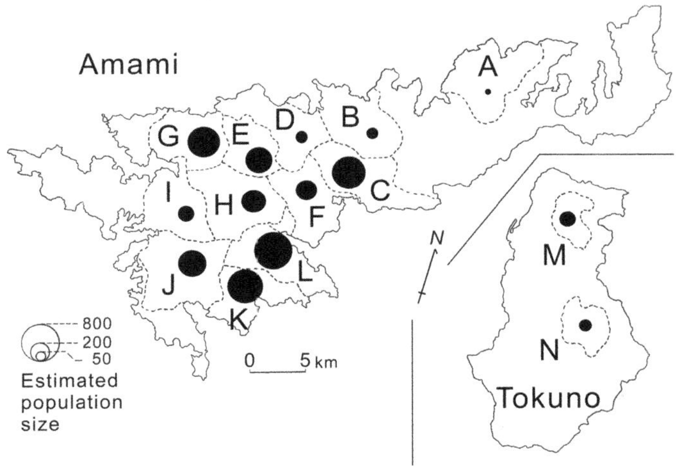

Figure 3.2 Geographical variation of the Amami rabbit population based on the estimation method described by Sugimura and Yamada (2004).

active from mid-March to late November, we conducted stream surveys from December to early March. We also counted pellets inside the forest to estimate pellet densities based on the following two correlations: (a) pellet numbers on the forest floor (pellets/hectare) and along streams (pellets/km), and (b) pellet numbers on the forest floor and along roads (pellets/km). The search technique on the forest floor used a line of 7–10 people separated by 1–1.5 m depending on the level of visibility.

We further divided the 14 regions using mountain ridges into 286 subregions with an average area of 212 hectare so that the regional variations in the rabbit's abundance could be analysed. The area of the rabbit's distribution was estimated based on the presence of pellets in each subregion. There were more pellets along the portions of forest roads that were close to streams than the portions further from streams (Sugimura, 1993). We concluded, therefore, that rabbits used areas near streams more heavily than other areas and that it was reasonable to estimate the local rabbit abundance based on catchment area as basic unit. We found rabbit faeces in piles comprising varying numbers of pellets (average of 29 with a SD of 22; Photo 3.1). It seems that a rabbit drops its pellets close to the ones it did within a few days so that we often found these latrines with a similar stage of oldness close together. Within the forests, we found an average of 0.11 pellets/m^2 over 12 sites. Averages of 0.47 and 0.97 pellets/m were found along the roads and streams, respectively. Sugimura *et al.* (2000) concluded that the total number of pellets/km along streams would be the most appropriate indicator of rabbit abundance. Sometimes it was very time-consuming work where the rabbit density appeared to be high.

We summed the areas of the subregions where we found pellets and estimated the distribution range: a total of 37,028 hectare on Amami and 3297 hectare

Photo 3.1 Piles of rabbit faecal pellets on a forest road.

on Tokuno Island. Regions A in Amami and M and N in Tokuno (Figure 3.2) were small and completely isolated from all the other regions by agricultural fields and residential areas. Regions K and L were also separated from the other regions by a major vehicular road with relatively heavy traffic. The size of these isolated populations would have been small relative to the total population of regions B–J. Pellets were absent from Amami's major peninsulas (forming parts of regions A, I and J). The number of pellets along survey routes had a considerable variance, which we thought may have been due to environmental factors. However, upon analysing the faecal pellet frequency in each subregion, I only found significant correlations between the frequencies of pellets in neighbouring subregions. This suggests that the movement of rabbits between subregions had a greater effect on pellet abundance than any environmental variable. Also, Sugimura et al. (2000) divided the rabbit's whole range into 39 areas by amalgamating some subregions and was able to obtain the following equation:

$$\ln P = 1.3 \ln N + 0.0019 * (68 - M) * M + 0.057L + 1.0 \ (R^2 = 0.61),$$

where P is the number of pellets observed/km, M is the percentage of mature forest, L is the percentage of cut-over forest, and N is the number of pellets observed/km in the neighbouring regions weighted by the length of the region's borders.

This equation suggests that the rabbit's abundance has a positive correlation with the percentage of mature forests. Thus, in general, the larger the area of mature forests, the greater is the rabbit's abundance. However, the equation also suggests that opening up a forest may contribute to an increase in the abundance of Amami rabbit. These results conform to the results mentioned earlier, in which a large number of pellets were sometimes found in areas where an old forest was adjacent to a clear-felled site.

A comparison of the results of this study with previous studies indicated that the Amami rabbit's total population had declined over the previous two decades. In 1977, Anon (1977) estimated the rabbit's range based on a questionnaire survey, interviews and faecal pellet searches on some forest roads during 1976 (Figure 3.3). Each day the researchers visited three roads each 1-km long and counted the number of new pellets. They reported that numerous pellets were found on every road visited throughout the rabbit's range. The study proposed that about 233 pellets/km/day was an average over the rabbit's distribution on the island, based on the above observations. Later, a series of studies in the central part of Amami Oshima (regions B–F in Figure 3.2) found that the number of faecal pellets on many roads fell to zero between 1976, the winter of 1985–1986 and the winter of 1990. The routes from which pellets disappeared accounted for 32% of the routes where pellets had been found before. The average faecal abundance decreased from 47 pellets/km/day in 1985–1986 to 13 pellets/km/day in 1989–1990 (Sugimura, 1987; Sugimura, 1993).

In a follow-up study (1993–1995), the average abundance within the whole range was 640 pellets/km for the same area. Using a conversion factor to make

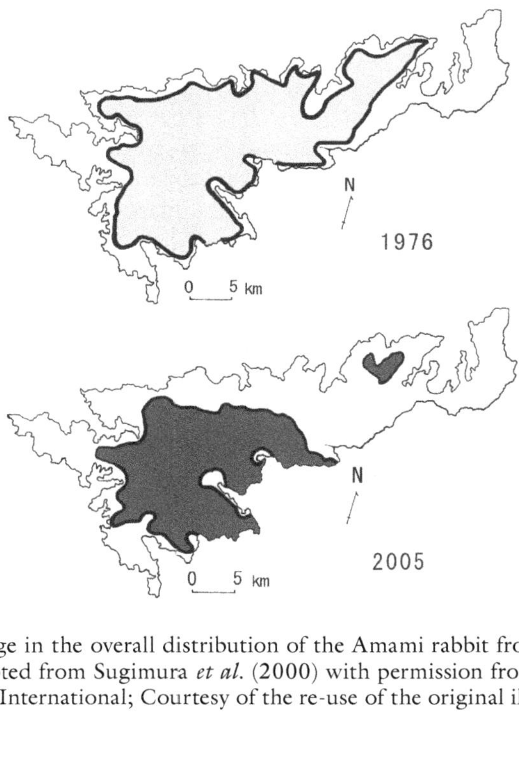

Figure 3.3 Change in the overall distribution of the Amami rabbit from 1976 to 2005. (Adapted from Sugimura *et al.* (2000) with permission from *Oryx*, Fauna & Flora International; Courtesy of the re-use of the original illustrations).

the units comparable, i.e., converting pellets/km/day to pellets/km or vice versa, Sugimura *et al.* (2000) concluded that the difference in abundance prior to 1986 and after 1989 meant that populations were declining. The decline in region B was most likely due to mongoose invasion. In contrast, in the other regions clear-cutting of mature forests was more likely to have caused the rabbit population decline. The study also suggested a remarkable fall in the rabbit's distribution. A comparison of the range estimated by Anon (1977) with the faecal distribution described in the 1993–1994 study (Figure 3.3) indicates that the rabbit's geographic range had indeed shrunk towards the central part of the island, thus isolating the population in region A from the others. However, the scale of the range reduction is uncertain because the methods for measuring the distribution range were different in the two studies.

Rabbits disappeared from parts of regions B and D, possibly due to the introduction of the mongooses, as described below. A substantial fall in population size must have occurred in region A, since Ikehara (1989) sighted a few individuals on a forest road in region A. Long-term field experience suggests that the rabbit population density is relatively high wherever they are sighted on roads. Much more intensive surveys in the 1990s and 2000s did not find any pellets on the forest roads in the same region. New roads were constructed in the time between the two survey periods, and other habitat disturbances, such as the development of a golf course, may have facilitated the population decline and the complete isolation of the populations in region A.

The Mexican volcano rabbit *Romerolagus diazi* is another important species in terms of lagomorph conservation (Chapman and Flux, 1990). Fa *et al.* (1992) and Velazquez and Heil (1996) indicated the importance of herbaceous cover in supporting high numbers of the volcano rabbit. In the case of the Amami rabbit, mature forests could be important habitat because they produce large amounts of acorns in late autumn as an essential food supply (Sugimura, 1987). The herbaceous cover was generally poor in old forests, except along streams and forest roads. Young forests and clear-felled sites have a richer herbaceous cover, and Sugimura (1987) observed numerous signs of feeding on perennial grasses *M. sinensis* and ferns *Dicranopteris pedata*. Kirino *et al.* (1984) indicated that rabbits in captivity were fond of *M. sinensis*, which was dominant in clear-felled sites and along roads in young forests, as well as acorns of *Castanopsis sieboldii*, which is the predominant tree species in mature forests (Shimizu *et al.*, 1988). Also, pellet abundance near mature forests was higher in winter than in summer (Sugimura, 1988). These observations suggest that the population density of Amami rabbits in mature forests may be more highly related to the level of acorn production than the herbaceous cover.

The 1993–1995 study indicated that although the Amami rabbit might not be on the brink of extinction, certain urgent measures needed to be taken if the species was to be conserved. The total rabbit population appeared to have been declining. The size of the geographic range of the Amami rabbit estimated was close to that of the volcano rabbit (280 km^2) (Hoth *et al.*, 1987; Fa and Bell, 1990). Conservation of the Amami rabbit might be regarded as urgent as was claimed for the volcano rabbit. Sugimura *et al.* (2000) suggested taking three conservation measures:

(a) Restrict extensive logging of mature forests to maintain a widespread mosaic of clear-felled and mature forests
(b) Control predators (feral cats, dogs and introduced mongooses)
(c) Halt forest road construction that may encourage further migration of predators through the forests

In the middle of the 1990s, extensive logging of mature forests stopped due to the change in the trade balance. So (a) was partially realised. It was volunteers who made the main contribution to (b) the mongoose control, and possibly other volunteer groups will succeed in sterilising most of the feral cats in the future. Measure (c) was achieved concurrently with (a).

On Tokuno Island where the two isolated populations might be critically endangered, large areas of forest have been replaced with farmland, so the two remaining ranges of rabbits were small. After we finished this study in 1995, another two groups conducted monitoring surveys using the same methodology. Both of them (unpublished) indicated that the northern population (region M) declined, while the population in the south (region N) increased. K. Shioya (Kagoshima-ken Kankyo Gijutsu-kyokai) told me that the total population probably had not declined significantly.

Sugimura and Yamada (2004) estimated the population size of the Amami rabbit based on faecal pellet numbers, the number of pellets produced in a day and the approximate age of the faecal pellets. Hayashi *et al.* (1984) gave the number of pellets produced in a day. To estimate the age of the pellet, I relied on my experience through 1985 and 1986, when I removed all the pellets from the survey routes and learned the characteristics of new ones that were no older than three days.

The basic form of the equation I used was $N = P_n/\ (A \times D)$ in which P_n is the density of new pellets (pellets/hectare), A is the average age (days) that the pellets maintain their freshness and D is the average number of pellets produced in a day by one rabbit (pellets/day/rabbit). A school in Amami once kept three rabbits, and Hayashi *et al.* (1984) reported the number of pellets produced by the rabbits over 30 days, which the school children counted daily. Consequently, I was able to estimate D at 147 (maximum 221 and minimum 62). Assuming a normal distribution curve with a sample size of 30, I estimated the mean, at the 95% probability interval, ranged from 130 to 164 pellets per night. At that time, the Natural Monument Act protected the Amami rabbit, and it was extremely difficult to obtain permission to catch any. I understand that the school was able to keep a few rabbits for many years because somebody with excellent trapping skills caught rabbits without a permit. Thus, this kind of valuable information resulted from possibly illegal activities.

We divided the known distribution of the rabbit into 12 regions, A to L, and further using mountain ridges into 124 subregions, which became the basic unit for estimating populations. Simply multiplying the density by the area of a subregion yielded a measure of the population size for each subregion, but the process was somewhat complicated, as shown in Figure 3.4. We counted pellets from February 2002 to August 2003 and compared the results with the pellet counts conducted from 1993 to 1994.

I estimated the approximate age of faecal pellets based on the presence of a sticky substance, which is a secretion from the caecum, and the glossiness of their surface. Then I classified the piles as either "new" or "old." The substance could be detected when "new" pellets were touched lightly, but it became dry and glossy when the pellets became dry and "old." I set up a survey route along a stream and monitored daily changes in pellet abundance and their external appearance over 15 days in 1997 and 1998. I removed old pellets each day and found a total of 96 piles of new pellets. The number of days that they retained the characteristics of being "new" was from one to five (mean 2.6; SD 1.1). The measurement of the pellets' age with this naïve technique was fuzzy, but the variance was small relative to the other essential components of the model (Figure 3.4). Because we were not able to collect sufficiently large amounts of data on the number of new pellets inside forests, we could not create regression equations with a reliable coefficient showing the correlation between the total number of pellets along streams and the density of new pellets on the forest floor (Sugimura and Yamada, 2004). In other words, estimates of the

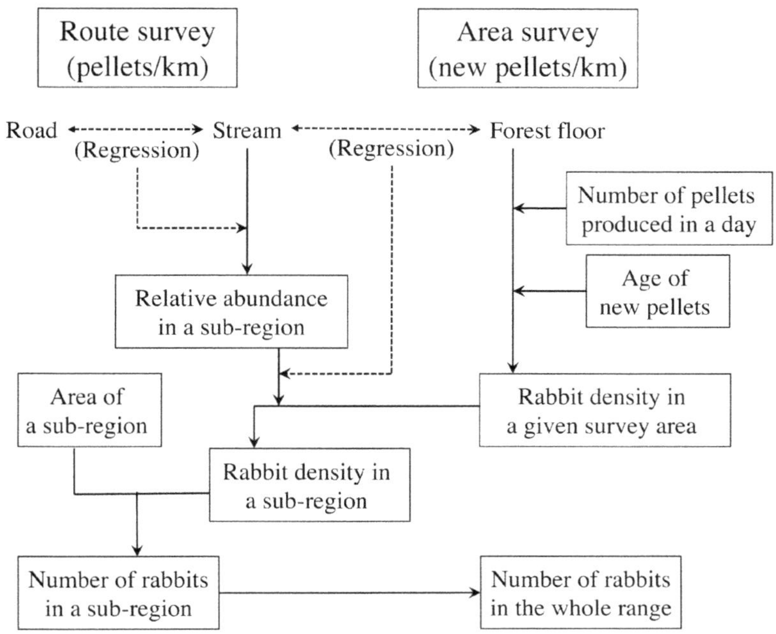

Figure 3.4 Estimation process of rabbit populations using the number of faecal pellets.

relative abundance of rabbits between the regions were less reliable than the rabbit density estimates inside forests.

Applying the field data and regression equations to the model (Figure 3.4), we estimated the rabbit population size in each subregion. This produced an estimate of the total number of rabbits at between 2500 and 6100 in 1993–1994 (Table 3.2). We found one of the two isolated populations in region A had disappeared by 2002–2003. The populations in areas of relatively high density, C, H, K and L, amounted to 47% of the rabbit's total population, while the area of these regions comprised 23% of the total. The estimated population decreased to 2200–5100 in 2002–2003, indicating a 15–18% decline over 9 years and the range also fell by 10%. The population decline was more evident in the northern parts of the distribution area—regions B, C and D. A comparison of these estimates with the number of rabbits sighted along roads at night supported the validity of the model (Sugimura and Yamada, 2004). Litvaitis *et al.* (1985), Krebs *et al.* (1986) and Murray *et al.* (2002) also indicated that numbers of faecal pellets were reliable estimators of lagomorph numbers.

A comparison between the 1993–1994 and 2003–2004 studies indicates the population trend between the two periods. The rabbit density declined in regions B, C and D (Table 3.2), where mongoose density was high (Ishii, 2003). In contrast, the rabbit density did not change significantly in the other regions, where mongoose density was relatively low (Ibid). As suggested previously, the rabbit population seems to have declined from the 1970s to the 1990s. This trend must have continued after the 1990s, mainly because of the mongoose invasion. The above model of population size enables a crude estimate of the population change (Figure 3.5). The population fall from 1976 to 1993–1994

Table 3.2 Estimated population size of the Amami rabbit

	1993–1994			*2002–2003*		
Region	*Distribution area (ha)*	*Density (/100 ha)*	*Population size*	*Distribution area (ha)*	*Density (/100 ha)*	*Population size*
A	771	1.1–2.5	8.3–19	599	2.6–6.1	16–36
B	1504	2.5–5.8	37–88	344	0.4–0.8	1.2–2.8
C	3050	12–29	360–880	3123	3.2–7.6	100–237
D	1501	4.7–11	70–165	1225	4.6–11	56–132
E	2403	10–24	245–576	2413	6.6–16	160–379
F	5273	2.8–6.8	149–360	5273	5.6–14	296–716
G	3549	8.5–20	301–722	3080	9.3–22	286–675
H	3022	7.3–17	221–525	3022	17–40	514–1207
I	2273	3.7–8.8	83–199	2458	3.2–7.5	79–186
J	2902	8.0–19	234–550	3426	6.5–15	224–526
K	2751	14–33	379–913	1921	6.5–16	125–302
L	3302	13–32	436–1062	3070	10–24	309–730
Total	32,299	7.5–19	2500–6100	29,953	7.2–17	2200–5100

The population size of a region is the sum of that in the subregions within each region. Therefore, the population size may differ from the product of area and density.

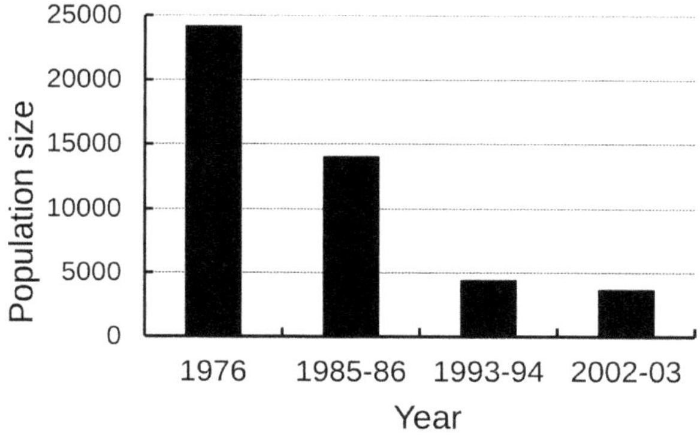

Figure 3.5 Crude estimates of the population size of the Amami rabbit from the 1970s to the 2000s.

was almost linear, while the decline was more gradual for the following 9 years. A couple of factors may have caused the latter, more gradual decline. First, clear-cutting forestry was practised all over the island before the early 1990s. In contrast, areas with a high mongoose density were limited to a certain distance from the original release point. Second, the regions where mongoose density was high—B and D in particular—had a lower rabbit density than most of the other regions (Figure 3.2; Table 3.2). Thus, considering all these factors, the overall impacts of logging may have been larger than those of the mongoose.

The collaborative work with the Amami Ornithologists Club from 2016 to 2018 was the last pellet count survey across the island. Despite limited time and budget, it revealed a crude trend for the island's rabbit population as a whole after the successful mongoose control. There was a considerable regional variation in the degree of recovery from the impacts of the mongoose invasion. The number of pellets counted per kilometre of road in regions E, F, G, H and J (Figure 3.2) from 2016 to 2018 was more than twice as many as that from 1992 to 1994. Near the East China Sea, there were no pellets on the roads from 1985 to 2003. Also, there was only a small number along streams in areas relatively close to the coast in 1993 and 2003, indicating that the population density there was low. The abundance along the roads in the same region in 2016 was 124 pellets/km. Thus, not only rabbit abundance but also its distribution range near the coast expanded to about 8 km west of the mongoose release point.

The number of pellets decreased by about half in regions B and C between 1992 and 2017. They were repeatedly found along a small path about 6 km southwest of the release point in 1985, while they could not be found for a distance of about 10 km away from the release point in 2017. In an area around 9 km south of the release point many pellets were consistently found until 2006, but three surveys conducted during 2007 and 2016 found none. Also, pellet numbers fell by about a third in regions K and L between 1992 and 2017, where the rabbit density was persistently high from 1992 to 2003. Considering all the changes in the regions, the total population size increased in recent years. However, the distribution range has not fully recovered from the impact of the mongoose.

There have been significant changes in road habitat conditions as feeding and pooping space during the period from 1985 to 2018. Roadside trees have become so tall that herbaceous plants do not grow well if a road is narrow and shaded. Some roads were paved and now attract more car traffic. Therefore, quantitative comparisons would not produce very reliable results on the degree of recovery. The MoE assessed the rabbit population size on the island, suggesting an increase of more than eightfold from 2003 to 2015, according to a newspaper report[3-2]. Although it used the same model as above, this population size appears to be an overestimation for several reasons. First, the number of survey routes was small, which could produce a large error range. There were more routes in the north near the East China Sea, where the rabbit population density was significantly lower than those to the south, as shown in the previous survey. A comparison between the 2016 and 2002 counts indicates that the rate of increase in this region was far larger than the others, possibly more than ten times larger. However, the numbers of pellets counted along the roads across the rabbit's entire distribution range do not suggest such a high level of population increase. The rabbit has two breeding seasons per year with one or two infants born per season (Suzuki and Oumi, 2017), so that its fertility rate is not very high. I think that the MoE should have used a meta-population dynamics simulation model for the assessment to validate its estimates.

3.3 Impacts of the mongoose on other species

Figure 3.6 compares the abundance of birds observed in 1993 with that in 2001. The mongoose population was not very large in the study area during 1993, and the population levels of the four species of birds were not significantly different between the two areas "High" and "Low" in the figure. Mongoose numbers exploded in the northern parts of the central mountains by 2001, causing significant differences in the mongoose abundance between the two areas. The numbers of purple jay and Ryukyu robin were considerably smaller in 2001 than 1993 in areas where mongoose density increased rapidly, i.e., the "High" area. The difference was not very large in areas where mongooses did not become so numerous, i.e., the "Low" area. The jay is found usually in the canopy layers, but parents leave juvenile birds on the ground, so mongooses would have easily been able to catch and eat them. Also, the robin is often on the forest floor to eat soil invertebrates. On the other hand, woodpeckers almost always live above the ground and no population decrease could be detected in 2001.

Based on night surveys, Watari *et al.* (2008) concluded that several species must have declined due to mongoose predation. They are the Amami rabbit, the ground nesting Amami woodcock *Scolopax mira*, the Ryukyu odd-tooth snake *Dinodon semicarinatum*, the Amami tip-nosed frog *Rana amamiensis*,

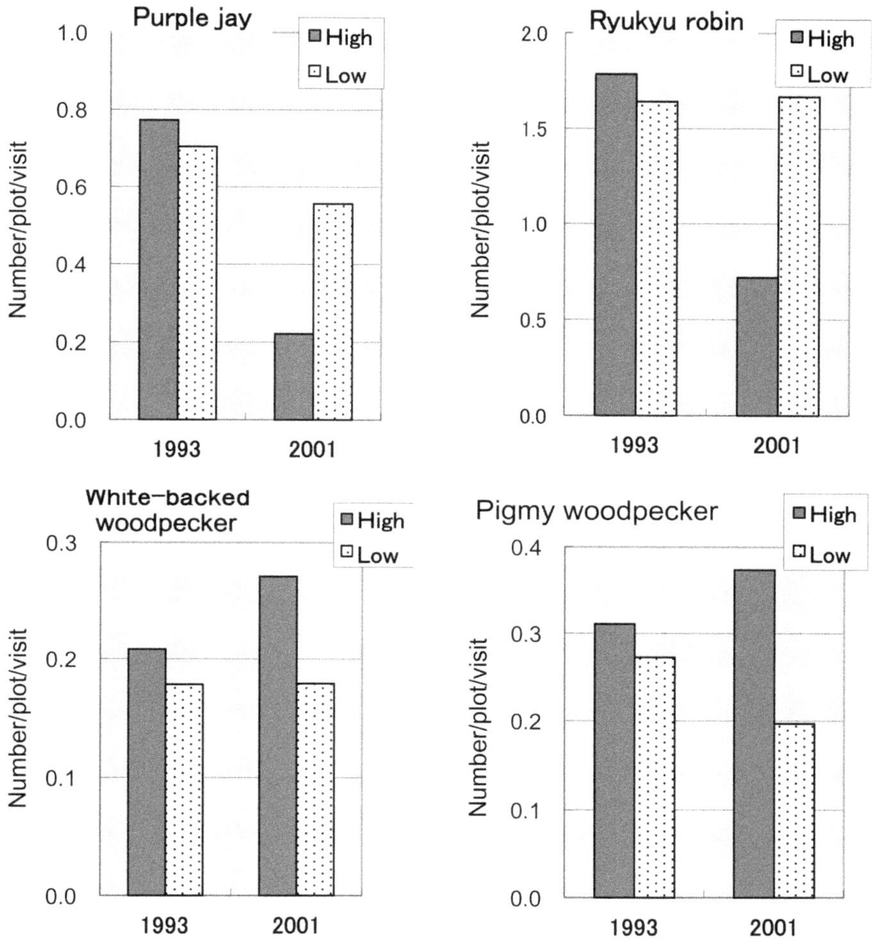

Figure 3.6 Comparison of the average number of birds observed at a sampling station per visit between before the mongoose population explosion (1993) and after (2001). "High" denotes the area where the mongoose density was significantly higher relative to the "Low" area in 2001. The "High" area was closer to the site where mongooses were released than the "Low" area.

the Otton frog *Rana subaspera*, the Ishikawa's frog *Rana ishikawae* and the grounddwelling Ryukyu short-legged skink *Ateuchosaurus pellopleurus.* This conclusion was based on the positive correlation between the number of animals counted and their distance from the mongoose release point. I partially agree with his conclusions, but the impact of the mongoose was not the only factor that contributed to such a correlation for some species, e.g., Amami rabbit and woodcock. Feral cats and dogs moved around the area of the release point before the mongoose invasion and could be expected to prey on native wildlife. So it is likely that even in 1985 the density of these native species already had some positive correlation with distance from the release point. There are two other species whose population declined considerably during the period from the 1990s to the 2010s—the Ryukyu green snake *Cycophiops semicarinatus* and the sword-tailed newt *Cynops ensicauda*. These are only a couple of examples among the numerous animal species that the mongooses could have caught on the ground or dug out from the earth—whether endangered or not, including mammals, birds, reptiles, amphibians and insects (Watari, 2008).

After the successful control of the mongoose, recovery has been gradual. Sugimura *et al.* (2013) selected six species that are likely to be adversely impacted to assess their recovery. They were the Amami rabbit, the Amami spiny rat, the purple jay, the Amami woodcock, the pale ouzel and the Ryukyu robin. We examined their rate of population change from 1985 to 2017 in the central mountains of the regions B–F (Figure 3.2). Generally, their numbers appear to have fallen sharply from 1993 to 2002 and recovered by 2009–2010 (Figure 3.7). Sugimura *et al.* (Ibid) calculated the Living Planet Index(3-3) to identify the general population trend of the six species, concluding that the populations had not returned to the 1985–1986 level by 2009–2010. This index measures the population trend of a given species group in an aggregate manner. The recovery of the six species was evident since 2003, as Watari *et al.* (2013) suggested a more than a fourfold increase in population levels using an index developed for the following species; the Amami rabbit, the Otton frog, the Amami tip-nosed frog

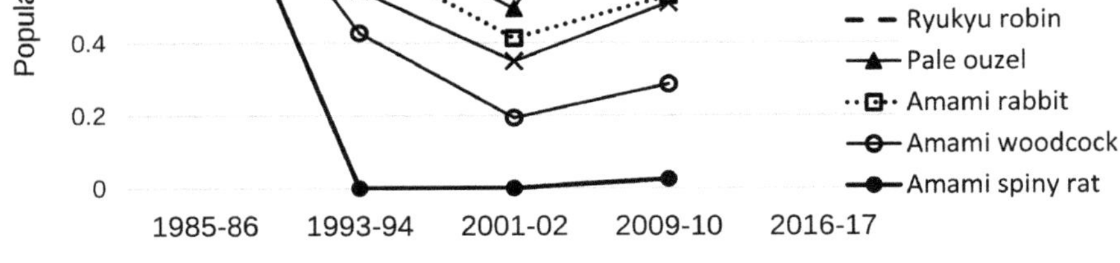

Figure 3.7 Trends in population size of wildlife species considered to be negatively impacted by the mongoose.

Population levels are shown relative to those in 1985–1986. Indices used are numbers/km (woodcock), faecal pellets/km (rabbit), numbers caught/trap-night (rats) and numbers/plot/count (the other species of birds) according to Sugimura *et al.* (2013). The lower the position in the diagram, the lower is the population level relative to 1985–1986. Rats and woodcock were not studied in 2016–2017.

and the Ishikawa's frog. The two studies surveyed areas where the mongoose impacts were severe, while I started my research in 1985 and Watari *et al.* started in 2003. The difference in conclusions is that I emphasise a slow recovery compared to the 1985–1986 population level, while the latter emphasised a rapid recovery rate since 2003 when the impact was most severe.

In the 1980s, it was almost impossible to obtain permission from the Agency for Cultural Affairs to catch spiny rats. However, the federal government gave researchers such approval for the first time in 1994. Unfortunately, they were not able to catch any in the areas where I had previously caught spiny rats. In 2004, I used the same traps and method as in Sugimura (1988). I caught many black rats *Rattus rattus* in regions B and C of Figure 3.2, but failed in catching any spiny rats, even in region E, where I did not catch black rats either. Due to the failure to catch a sufficient number of spiny rats for quantitative analyses in the period of 1993 to 1994, I can only think that the population must have decreased considerably. Although the two species of rats are nocturnal, the apparent complete disappearance of spiny rats from the study area in 1985 and 1986 suggests a great adverse impact from mongoose predation.

Forest logging had more apparent adverse effects on the robin than the other species in Figure 3.7, as indicated in the previous chapter, so I suppose the two species of mammals, the jay and the woodcock had similar paths of population change from 1985 to 2010. As at 2017, I do not have any data on the spiny rat and the woodcock. As discussed above, the spiny rat may well have been able to coexist with the poisonous snake, the habu, but was vulnerable to mongoose predation. In fall, rats depend on acorns for food. The *Castanopsis* trees have grown older and Figure 2.9 suggests that on average they currently produce more acorns than some years ago.

It will take years but eventually the rabbits, woodcocks and spiny rats will probably recover their population levels to those of 1985–1986. The ouzels are winter birds that come to Amami in large numbers and Watari (2008) concluded that they might be the most critical winter food for the mongoose. Because they migrate in considerable numbers to south-western Japan, the impact of mongoose predation would have had minimal effects on the species survival as a whole. The jay population appeared to have recovered so well that the MoE removed it from the national Red List. Another species that increased its numbers is the Amami thrush that was scarce in the study area during the 1980s (Chapter 2). Its population was already small in regions where the mongoose density became very high, so it may not have been greatly impacted by the mongoose. However, monitoring of its populations did not cover the southern parts of the island. Although the process of population increase is not understood, the species is currently abundant even near residential areas of Amami City. On the other hand, the robin population seems to have been declining recently, despite forest recovery and successful mongoose control.

Recently, researchers have been paying more attention to the potential impacts of stray cats because automatic flash cameras have recorded the presence of many cats in the forests in the interior parts of the island. Another positive step is the progress made by local volunteers towards the protection of rare species by introducing community education programs and regulations for pet sterilisation.

3.4 Towards proper forest management

Forest managers used to perceive Amami's forests primarily as land for growing wood products. As a result, they attempted to log them as often as possible. However, imported wood outcompeted and stopped the clear-cutting of the last remnant of old-growth forests. It was not conservation efforts that saved the island's unique biodiversity but the changing global economy during the 1990s. Then a turning point came when the MoE began its attempts to eradicate the invasive mongoose in 2000. It also opened Amami Wildlife Centre with an exhibition hall, which had about 15,000 visitors in the same year, although the number of visitors subsequently declined. The ministry decided to propose the forests of the Nansei Archipelago for UNESCO World Natural Heritage status. The first proposal was not accepted and it is waiting for a response to its revised proposal. Amami airport is full of tourists in recent years, who visit the island to see its unique species of plants and animals. The number of companies running ecotourism businesses increased from zero before 1990 to 39 in 2016, while many more rental cars are seen in forested areas. Another great advance is an establishment of Amami Gunto (island group) National Park, as described below.

One of the significant differences from the 1980s and the 1990s is the emergence of nature tourism in the island. There were 30 forest tour courses in 2016, 28 of which included programs to show and explain plant and animal species at a total of 42 sites (Sugimura, 2016). The most popular place is Kinsakubaru, which is less than an hour from Naze, where most of tourists stay overnight. The area is only 125 hectare, but it is one of the three remnant national forests that forestry records show that they have been uncut for more than one hundred years. It does not have many large trees since it is located near a ridge with relatively infertile soil and is vulnerable to strong typhoon winds. M. Tabata, a local amateur botanist, considers that the national forests lower in a valley had a better habitat for unique species. However, the forests here were clear-cut during the 1970s and the 1980s because they had a greater biomass per unit area. At that time, a forester working for Amami National Forest sent a letter to the agency director in Tokyo asking that the plan be withdrawn so one of the last remnants of old forests could be saved. We don't know how it was discussed, but the final decision was to leave it uncut and protect it as a recreation forest with a good quality forest road. Only a small number of visitors came before 2000, but a camera placed in the forest photographed a total of 1330 people in August 2016 with 85 as the daily maximum in October (Ibid).

Night tours are also popular, in particular, the "Amami rabbit watching tour," even though it has a few weaknesses. The tour carries customers in a vehicle (Photo 3.2). There are not many roads along which it is likely to see a rabbit any day. There is only one road that is nearly always full of new faecal pellets; it

Photo 3.2 Amami rabbit watching tour.

is paved and not very far from Naze. A remote camera showed a much smaller number of cars there than the number of visitors in Kinsakubaru (Figure 3.8). However, a total of 195 vehicles visited in August 2016, which may have significant impacts on the rabbit populations. There is another piece of evidence that revealed the popularity of forest tours. That is an interview survey conducted at Amami's airport, in which 14% of the 122 respondents answered that they joined a forest tour. Times have changed from the days when administrative officials and politicians argued against conservationists, by saying that nature conservation would bring no monetary benefits to the local communities in contrast to large government subsidies.

The MoE patiently continued its efforts to establish a new national park, and one of its challenges is to prevent severe impacts from forestry. As discussed earlier, clear-cutting for woodchip production is barely economically sustainable so loggers cannot economically cut forests far from a mill. After all the mills were closed down in the 1990s, Uken Village built one and Iwasaki Corp. built another in 2007. This new investment reflected an increase in the price of import wood and the growth of trees in forests cut during the 1960s. The ministry had to negotiate with the private company over national park zoning because Japanese national parks can include large areas of private land after agreement with landowners. Zoning is essential as a matter of species habitat conservation. Now that the company owns the largest and most important parts of the habitat areas in the central mountains, negotiation with the company was undoubtedly critical to its success. Commercial harvesting is not allowed once the company agrees that its land can be allocated as a special protection zone or a first-class special zone. However, the ministry did not have sufficient funds to match

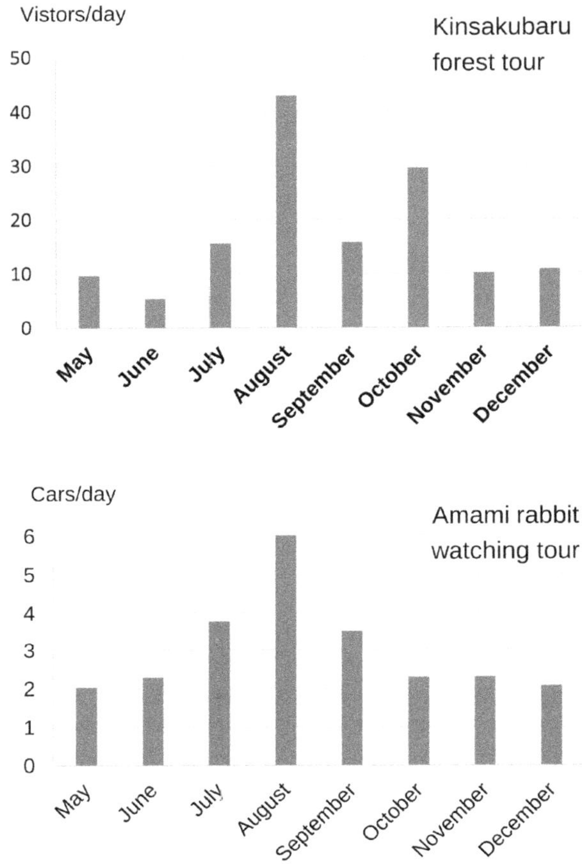

Figure 3.8 Number of visitors and vehicles in forest tours.

the company's request. My interviews with officials at the local wildlife centre touched on the difficulties inherent in the negotiations to establish the park. During the last set of negotiations, senior staff from Tokyo had to be involved. It was therefore surprising that the company's president decided to completely withdraw the company from forestry in 2017 and agreed that the park would incorporate large parts of the company's land into its core areas. The major reason for this must have been that forestry was no longer profitable.

These changes allowed the development of a new forest management structure containing a reasonable balance between forestry, species conservation and tourism. A crucial component is the national park zoning. During the period from 2012 to 2017, clear-felled sites were less than 0.2% of the forested land in total (Table 3.3) and located in the west, outside the core parts of the unique species' distribution areas. The private company cuts trees only in Yamato

Table 3.3 Area of logged sites in each municipality of Amami Oshima (ha)

	Yamato	*Uken*	*Setouchi*	*Amami*	*Tatsugo*	*Total*
2012	55.7	30.8	1.5	10.9	0.0	99.0
2013	38.9	8.5	6.2	13.0	0.0	66.5
2014	28.7	38.7	3.6	9.8	0.2	81.0
2015	19.8	26.0	1.6	2.6	0.0	50.1
2016	1.0	12.8	10.3	1.6	1.1	26.8
2017	21.2	22.4	18.6	0.0	0.0	62.3

Village and, as mentioned, was not going to continue its forestry operations. The national Forestry Agency decided that it would protect most of the forests as a "forest ecosystem preservation area," in which national law requires that loggers must never cut trees. The rest of the area is too small for economical forestry operations, but might be cut for other reasons. Summing up, the national park has strictly protected areas as special protection zones and first-class special zones covering about 12,000 hectare, or 37%, of forests. The Japanese Natural Park Law specifies arbitrarily that natural ecosystems should be preserved in harmony or coordination with forestry activities in the second-class special zones, but not in practical terms how they can be achieved. Japanese love the expression, "in harmony with nature," but what does it mean in reality or in practice? Different interests can interpret it differently, so there were arguments during the meetings of stakeholders such as the directors or officers of the MoE, Forestry Agency, local municipalities and Kagoshima Prefecture, as well as loggers and chip mill managers. They eventually came to an agreement that allowed a total of 10 hectare clear-cuts in a given forested watershed areas. The accepted practice on Japan's main islands of Kyushu and Honshu allows 5 hectare clear-cuts in second-class special zones. An officer in Uken Village told me that the reason the area was larger in Amami was that the central office of the MoE in Tokyo declared that having an agreement with foresters should be one of the feature points in the proposal for the UNESCO World Natural Heritage Area. There were several conditions included in this agreement, such as leaving narrow belts of uncut trees, 20-m width on each side, along streams and ridges, and 20 m in width outside the strictly protected zones. Cutting forests over 80 years old and those on slopes over 30 degrees is prohibited. A rough estimate is that about 10% of the forests would be left uncut in a watershed (Photo 3.3). These decisions ignored the studies on the adverse effects of clear-cutting on wildlife species.

Areas of strictly protected forests are limited to about one-fifth of the forest land on the island. Loggers will cut the western parts of Uken Village and Setouchi Town shortly, but fortunately, the areas are geographically separated from the areas popular for forest tourism (Figure 3.9). As already discussed, forest logging has significant effects on some wildlife species, and although

Photo 3.3 Clear-cutting in the second-class special zone in Amami-gunto National Park.

forests recover over a number of years, just how long it takes depends on site conditions. Satisfying the habitat requirements of each species should also be of concern. There are two conditions necessary to create suitable wildlife habitats, the production of acorns and the formation of cavities in tree trunks, both of which are roughly correlated with forest age. Acorns are essential food for such species as the Amami rabbit, spiny rat, long-haired rat, purple jay and wild

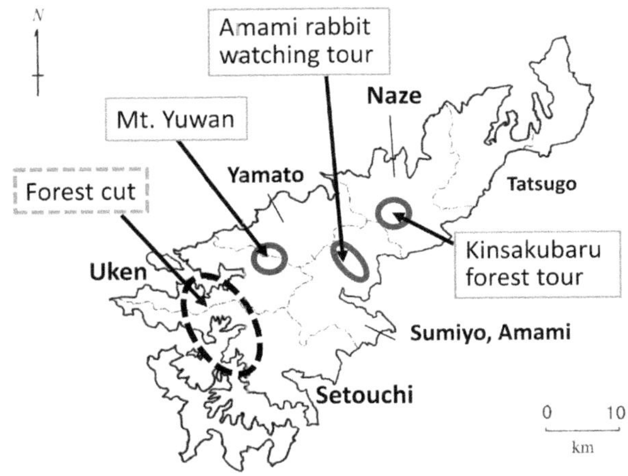

Figure 3.9 Locations of popular forest tour sites and the area in which forests were recently logged and are expected to be logged in the near future.

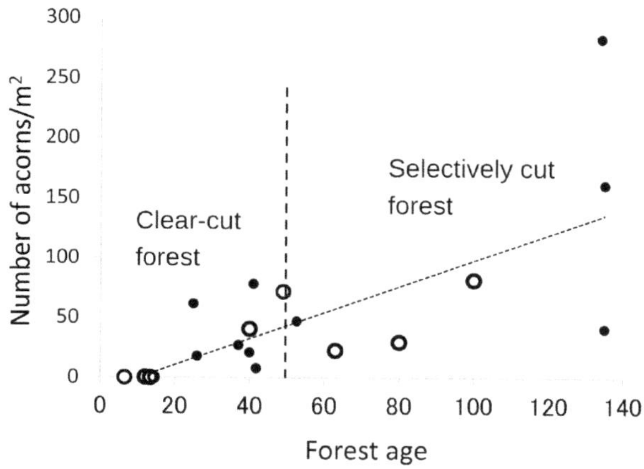

Figure 3.10 Relationship between forest age and the number of acorns of *Castanopsis sieboldii* on the forest floor. *White circles* are acorn densities as surveyed in 1985–1986. Solid black dots are densities as surveyed in 2010. Forests no younger than 40 years of age were not clear-cut and had relatively large-diameter trees in 1985–1986, while such forests grew to 65 years old and over in 2010.

boar. Young forests yield few acorns, if they are less than 20 years old, while acorn productivity does not correlate with forest age among older forests (Figure 3.10). However, on average, forests over 50 years old produce more acorns than younger forests. On the other hand, large-diameter trees provide nest sites for the long-haired rat, Ryukyu robin, woodpeckers and two *Parus* species. The long-haired rat is the scarcest among them. I measured the diameter at breast height (DBH) of some canopy trees in the forests where the rats lived and found that the average DBH was over 30 cm. Only forests over 80 years old contained sampled trees with 30 cm or more as the average DBH (Figure 3.11). An additional habitat condition is the presence of herbaceous plants for Amami rabbits.

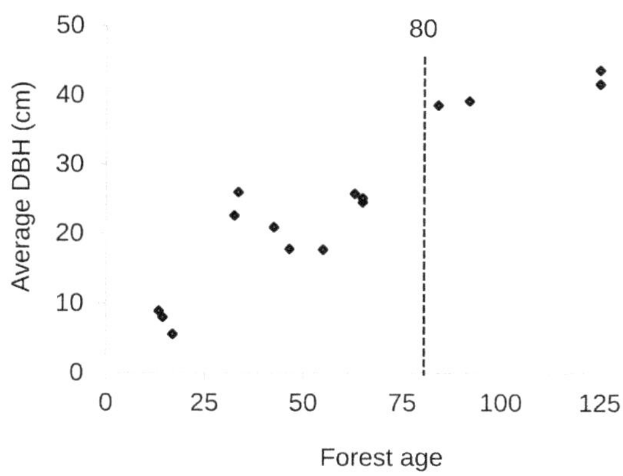

Figure 3.11 Relationship between forest age and the average DBH of canopy trees.

They are more abundant in logged sites and there was a positive correlation between the number of faecal pellets found on survey routes and the area of recently cleared forest where tall grass *M. sinensis* dominated the regenerating trees (Sugimura *et al.*, 2000). How long such a state lasts before forest recovery depends on a variety of conditions, such as the dominant tree species and the growth of *M. sinensis*. Slope orientation also matters because winter winds inhibit forest recovery. According to my observations, it took about 15 years at the longest before a dense stand of trees with closed canopies took over clear-felled sites. In summary, forests containing the necessary conditions for the main local wildlife species would be forests over 50 years old for acorn production, forests over 80 years for cavity formation and forests younger than ten years for herbaceous cover.

The last stage of my studies on Amami was looking into the future. I conducted an integrated forest management analysis that used forest age, ownership and national park zoning. First, it evaluated habitat conditions of the Amami rabbit as indicator species for animal watching and long-haired rat as an indicator for vulnerability to logging. Forest age also served as an indicator for primeval forest tours and bird watching. Rotation cycle determines the forest age structure over the long run, and the managers of Amami Gunto National Park determine rotation times in the three zoning categories; core, buffer and outside the park. Core areas include special protection zones and first-class special zones, in which logging is prohibited. Buffer areas are the same as second-class special zones because there are no third-class zones in Amami. The current plan in the buffer areas is a short rotation cycle with some restrictions as discussed above. Long rotation cycles could be another option, but it is ignored by the MoE and by the Forestry Agency. The forests in the third category, outside the park, are managed on short rotations with no restrictions; they are regarded primarily as an exploitable resource. So, the options I chose for rotation cycles were 50 years representing the current state in buffer zones and outside the park, and because buffer zones are supposed to contain long and very long rotation cycles, 100 years as a long rotation and 200 years as a very long rotation cycle. By integrating the two dimensions, area zoning (spatial) and cycle length in each zone (temporal), I proposed six potential scenarios (Table 3.4).

Table 3.4 Forest management scenarios (A to F) under a variety of rotation cycles (years)

	Area (ha)	*(A)*	*(B)*	*(C)*	*(D)*	*(E)*	*(F)*
Core	11,541	No cut	No cut	No cut	No cut	No cut	No cut
Buffer	22,161	50	100	200	No cut	No cut	No cut
Outside park	14,556	50	50	50	50	100	200

Scenario (A) represents the current state in which the rotation cycle is 50 years. The other cycles have not been realised because of conditions described in the text, but the other scenarios supposed 100 years as a long rotation and 200 years as a very long rotation cycle.

Table 3.5 Ultimate forest age composition produced by adopting the scenarios A to F in Table 3.4, based on two logging area scenarios

	Restricted to Uken & Setouchi					*All municipalities*				
	0–10	*11–50*	*51–80*	*81+*	*Area cut/yr*	*0–10*	*11–50*	*51–80*	*81+*	*Area cut/yr*
A	2.9%	11 %	0%	86%	197 ha	14%	57%	0%	29%	690 ha
B	2.3%	9.3%	1.6%	87%	159 ha	10%	41%	12%	37%	491 ha
C	2.0%	8.2%	0.8%	89%	140 ha	8%	32%	6%	53%	391 ha
D	1.8%	7.1%	0%	91%	121 ha	6%	24%	0%	70%	291 ha
E	0.9%	3.5%	2.7%	93%	61 ha	3%	12%	9%	76%	146 ha
F	0.4%	1.8%	1.3%	96%	30 ha	2%	6%	5%	88%	73 ha
P	7.6%	59%	25%	8.3%	40–80 ha	Same as Case (1) on the left				

Assumed cases: (1) Logging is restricted to two municipalities, (2) Logging of every unprotected forest. Refer to the text to see the reasons for choosing these two extreme cases.

P shows the current age class structure.

The next step was applying these scenarios to forested areas to identify the future forest age composition under each scenario, when the age composition would eventually stabilise, i.e., at around 80 years. I use two options as to areas where logging is possible. One assumption allows no new chip mills and restricts logging to two municipalities, Uken and Setouchi. The mill in Yamato was closed down in 2018. The other option allows new mills and assumes that logging would be economically feasible in every forest based on an increase in the demand for woodchips. Incorporating these two options produced 12 scenarios and simulation outcomes, as shown in Table 3.5. The scenarios have a few notable features. First, a large portion of the forest area would be maintained as a suitable habitat for those species that require large trees as long as the cutting area is restricted to Uken and Setouchi. Second, there is a wide range in both the area of forest logged each year and the percentage of remnant old forests. Nevertheless, it is unlikely that loggers will cut all the forests in the buffer and outside the park as in scenario (2)-A. This is because it is difficult for loggers to contact all of the many small landowners and because some local communities will not permit logging so as to protect the upstream watershed and landscape (H. Yamashita (Uken Chip Mill), pers. comm.). Third, the area of old forests over 80 years will increase to be at least three times the current area, which guarantees that habitat conditions will be greatly improved for species such as the long-haired rat. Fourth, a decline in the area of clear-cutting may not be favourable to the Amami rabbit, as Sugimura (1988, 2000) suggested.

Thus, not all conditions would be favourable for the species of conservation concern. We also need to consider a couple of recent developments. One is that the Forestry Agency started a subsidised project, in which local forest workers remove shrubs and trees in the forests' middle and lower layers. The effects of such operations on the forest structure are apparent (Photo 3.4). Although

Photo 3.4 Forestry Agency's project to enhance growth of young trees by removing shrubs and small trees in the middle and lower layer (Above: after the treatment; Below: before)

ostensibly promoting carbon sequestration and the growth of useful trees, its real purpose is no doubt to create employment and secure subsidies from the national government. Because the workers leave logs and twigs on the forest floor and let them decay, managing forests in such a way would be expected to produce more carbon emissions than leaving the forests unmanaged. Large areas of forests have already had their vertical structure dramatically altered without paying any consideration of the impacts on the local forest ecosystem and biodiversity. Okinawa Prefecture has conducted the same kind of activity for a few decades, but the only commercial use it found for the trees was chip wood. A prefecture government official told me that it had been a "subsidy project" to meet the demand for employment, and this was also true for Amami. The other important matter is a rapid increase in the demand for forest biomass energy in Kyushu. Kuboyama (2018) expected that this demand would soon double. Strictly protected areas on Amami are not large even after the creation of the park. The areas cut before the 1970s are now ready to be logged again. There were two separate plans to build a new chip mill through the last decade, but they were not realised partially due to the opposition by local citizen movements. The bad news is that FSC Japan decided in 2018 to approve certification to the wood produced from Amami Gunto National Park (C. Mitsushiba, pers. comm.). I made a submission opposing this decision to Japan's FSC, but it did not act on my criticisms and suggestions.

This chapter and Chapter 2 describe the changes in the significant factors and stakeholders driving forest use and management on Amami (Figure 3.12). Until the early 1990s, government subsidies supported forestry that attempted to generate an economic profit, while there was no consideration of conservation values by the local governments. There were no strong and effective actors or drivers for alternative forest management perspectives through the 1990s and the early 2000s. Domestic forestry was outcompeted by cheap imported wood, while on the global level, concerns for nature conservation were growing after the 1992 Rio Summit. Although scientists warned of the impacts of invasive mongoose on Amami's unique biota and began their own voluntary research activities, the government did not take action until 2000. After the

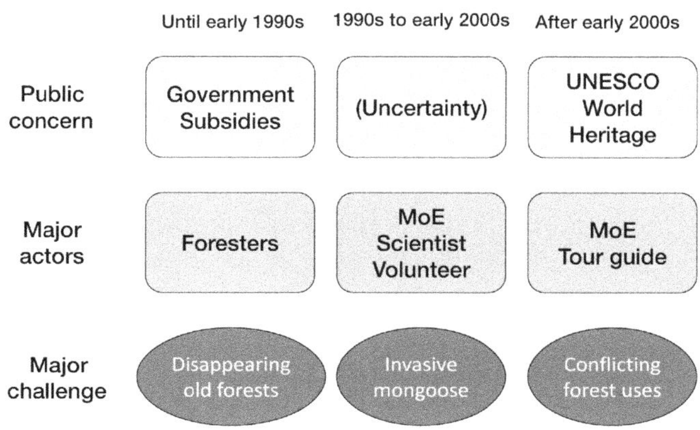

Figure 3.12 Changes in major public concerns, actors and challenges of forest use and management on Amami Oshima from the 1990s through the 2000s.

bubble economy collapsed in the early 1990s, the society seemed to have lost a kind of common sense, mammonism; reliance on coinage or on monetary values in the pursuit of happiness (Ueda *et al.*, 1993; Sanbe, 2018). In 2003, the MoE proposed Amami for UNESCO World Natural Heritage status, and biodiversity conservation and nature tourism gradually became of significant concern on the island. The application process meant that large areas of forests were designated as a national park, while leaving room to revive forestry. The current state is much better than the past, but future conservation of Amami is not yet assured. Critical decisions in the decision-making processes used to determine the national park zoning, forestry operations and FSC certification, were basically "top-down." Relevant decision-makers did not seem to have sufficient local experience and knowledge to understand the complexity of the situation on Amami. In reality, how wildlife and the local society interact on Amami is influenced by biological, silvicultural, social, geographical and historical factors amongst others. It is this interaction that generates the way the island's natural landscapes are modified by humans.

Supplementary notes

3-1) *Heisei 30-nendo Amami Ōshima niokeru mangūsu bōjo jigyō no jisshi kekka oyobi reiwa ganendo keikaku nitsuite.* Available at: http://kyushu.env.go.jp/okinawa/pre_2019/post_84.html. (accessed 28.05.20).

3-2) *Zetsumetsu kigu amaminokurousagi, zōka ka, tenteki kujo ga sōkou.* Available at: https://www.asahi.com/articles/ASM2M67FTM2MUTFL016.html. (accessed 16.05.20).

3-3) The Living Planet Index selects a given group of species, assesses the rate of change in each population of a species and then averages the rates over time to measure aggregate population trends of a species group, such as birds, herbivores or arboreal species. The rate of population change, dt, is calculated as $dt = \log(Nt2/Nt1)$ where $Nt1$ and $Nt2$ are the population indices at time points, $t1$ and $t2$, respectively. In Figure 3.7, this dt is set to 1 at the initial time point, 1985–1986. Loh *et al.* (2005) describes details of calculation of the Living Planet Index. It is particularly useful where abundance has been measured in different ways for different species (e.g., frequency of observation, trapping, field signs) (Sugimura *et al.*, 2013).

4 Forest utilisation in Fukushima before and after the 2011 power plant accident

Introduction

In 2011, Fukushima suddenly became famous worldwide because of the Fukushima Daiichi Nuclear Power Plant accident. Before the accident, it was the most popular destination among Japan's 47 prefectures for those migrating from urban areas to enjoy rural life. Many Japanese live far from Fukushima and they were not able to grasp the geographical range and the levels of contamination very well. They were reluctant to purchase any food produced in the prefecture as well as to visit there, and they probably did so, because of the impression they received from the name "Fukushima Daiichi" (Daiichi means the first). The power plant was on the east coast facing the Pacific Ocean. The western part of the prefecture was much less contaminated than the northern part of Kanto region, located south of Fukushima and north of Tokyo. The higher pattern of contamination in the Kanto region was the result of a change in wind direction which initially blew northwards but change to southward a few days after the accident.

Soon after the accident, the Ministry of the Environment started measuring radioactivity and decided to remove the surface layer of soil in residential areas and farmlands with radioactive caesium (Cs) in the soil exceeding 5000 Bq/kg in Fukushima Prefecture. This was done because radioactive caesium had adhered firmly to fine-grained clay and would emit radiation for an extended period. Forests were neglected unless they were less than 20 m away from human residences, crop fields and vehicle roads. The ministry did not realise the importance of forests to local people and concluded that they were too large to remove all the surface soil. Reflecting the opinions received from local people, Fukushima Prefecture government protested the ministry's conclusion, but the ministry did not change its decision without verifying its feasibility.

During the 4 years prior to the accident, I visited the western part of Fukushima many times and the eastern part occasionally to develop quantitative methods to evaluate forest ecosystem services. I operated a research project, in which biologists, economists, geographers and local non-governmental organisations (NGOs) participated. On 11 March 2011, when the Great East Japan Earthquake occurred, I was in Sapporo, Hokkaido, with my colleagues to present some of the final results of the project. After the accident, scientists issued many reports on various aspects of ecosystem state and function but few from the standpoint of forest ecosystem services except for wood production. I felt obliged to assess any significant effects of the accident on forest ecosystem services and started a few small-scale projects after I returned from Indonesia in 2013. This chapter has three sections which describe the following topics:

(1) Wildlife species diversity in comparison with three other regions, Amami, Tsukuba (satochi-satoyama landscapes) and northern Kyushu
(2) Importance of forest ecosystem services for local people and visitors focusing on Tadami Town, on the southwestern end of the prefecture
(3) Changes in the ecosystem services after the accident

4.1 Geography, beech forest ecosystems and wildlife in Tadami Town

4.1.1 *Natural geography of Fukushima and Tadami town*

Fukushima is approximately 200 km north of Tokyo (Figure 1.1), ranking 3rd by area and 14th by population among Japan's 47 prefectures. It is one of the six prefectures in the Tohoku region, located in Tohoku's south end and north of Kanto. This location is economically advantageous since the Kanto region has the largest population of any region, accounting for about one-third of the country's people. Fukushima's forested area is 9720 km^2 covering 71% of the land. Geographical variations are generally greater in the east-west than the north-south direction. Two mountain ranges, extending north-south, Abukuma in the east and Ohwu in the centre, divide the prefecture into three regions: Hama-douri, Naka-douri and Aizu (Figure 4.1). These geographical divisions are commonly used in the prefecture, accounting for similarities in climate, terrain, population distribution and main industries within each region. The two mountain ranges contrast each other. The Abukuma Mountains are relatively low and gently sloped; the highest peak is Mt Ohtakine (1192 m). The highest peak of the Ohwu Mountains, Mt Azuma (2035 m), is about twice as high. The Abukuma Mountains are geologically old and non-volcanic, while the latter has active volcanoes. Douri means a street, reflecting that each of the two regions has a main route of transportation connecting cities of commercial and administrative importance in the north-south direction. Hama and Naka mean coast and centre, respectively. Hama-douri has small plains along the Pacific Coast gently sloping upward to the Abukuma Mountains. One of the unique terrain features of the Abukuma is that there are many basins and rural settlements further inland beyond the relatively steep slopes with narrow valleys near the coast. Such settlement locations are distinct characteristics of geologically old mountain ranges, which are typically observed only in two regions in Japan (the other is the Chugoku Mountains).

The climate of Hama-douri has the smallest seasonal and daily temperature range, the warmest winter and the least amount of snow among the three regions of Fukushima. Naka-douri has the largest plain in the prefecture along the Abukuma River, as well as the largest population and the best transportation access from the Tokyo metropolitan area. Aizu is characterised by one of the areas

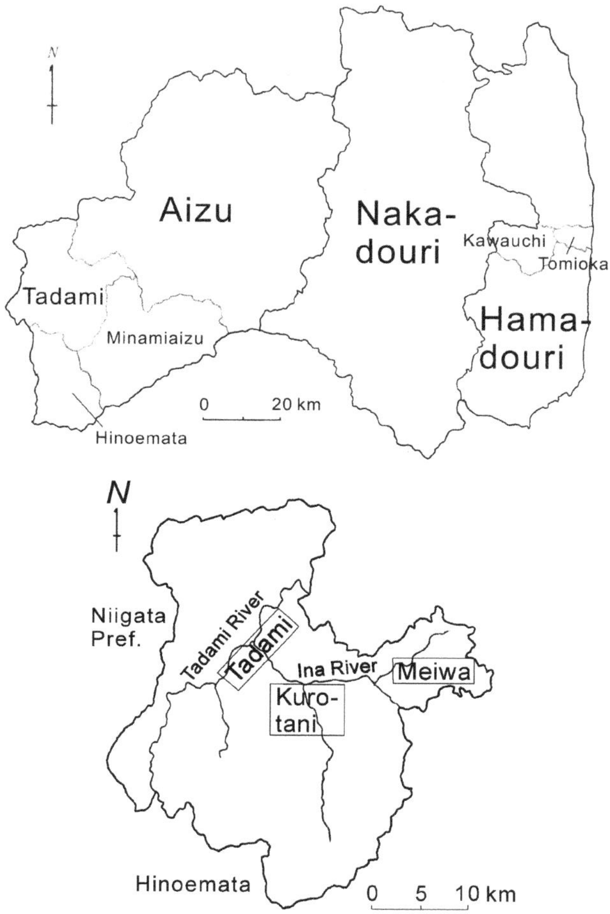

Figure 4.1 Three geographical divisions of Fukushima Prefecture and municipalities surveyed in the project (*above*) and the locations of the residential districts and rivers in Tadami Town (*below*).

with the heaviest snowfall in the country. It has the smallest population, distributed along narrow valleys and small basins, and has the worst access from Tokyo. Another distinguishing difference among the three regions is the proportion of tree plantations relative to natural forests. Tree plantation cover was 61, 44 and 20% for Hama-douri, Naka-douri and Aizu, respectively in 2017. The reason why it is so low in Aizu is mainly due to heavy snow. Hinoki cypress *Chamaecyparis obtusa* is unable to survive winter with deep snow. Sugi cedar *Cryptomeria japonica* does not grow straight near the base (Photo 4.1), shaped by the downslope pressure of accumulated snow therefore, the wood quality is not very good.

Tadami Town (hereafter "Tadami") is in the southwest of Aizu, and it has attempted to take full advantage of large areas of natural beech forests to

Photo 4.1 Many sugi cedars *Cryptomeria japonica* do not grow straight near the base due to heavy snowfall.

sustain its population and economy. It is located at 37 degrees North latitude and 139 degrees East longitudes, about 10 degrees north and east of Amami Oshima. It has very distinct seasons. The local meteorological weather station located at 377 m above the sea level reported the average temperature of the coldest month (January) as –0.9 degree Celsius, the warmest month (August) 24.6 degree Celsius and the annual precipitation 1893 mm in 2019. The leaves of the beech sprout light green in May and the fresh green leaves gradually appear at increasing elevations along the mountain slopes, as the days go by in late May and June, towards the end of forest roads at an altitude of 1000 m or more. Another attractive season to visit is late October, when the beech leaves, among others, display autumn colours. Also in the fall, many local people are busy picking a variety of mushroom species mainly for personal consumption. The town is surrounded by high mountains around 1500 m high in the west bordering with Niigata Prefecture, another range of mountains from 1500 to

1800 m high in the south and the third range around 1000 m high in the north and east. Water flows down from these mountains into two rivers, Ina and Tadami, along which are most of the residential areas (Figure 4.1). Prevailing northwest winter winds bring heavy snowfall after humid air coming over the Sea of Japan contact the mountains bordering with Niigata Prefecture. Snow depth occasionally reached 10–15 m. The amount of precipitation is the largest in winter, and the rivers carry snowmelt water from the mountains in spring and summer. Although winter is the most challenging season for the residents of Tadami, the average temperature in January increased from –2.2 degree Celsius during the decade 1980–1989 to –1.3 degrees during 2010–2019. The lowest temperature recorded during the same periods also increased from –12.2 to –10.8, probably due to the effects of global warming.

4.1.2 *Society and land use of Tadami*

The area of Tadami is 748 km^2 and the population was 4178 in October 2019. Population decline is a severe problem for sustaining many of the 42 residential settlements within the town, and I will discuss this issue at the end of this chapter. According to the statistical handbook of Tadami Town, there were 26 villages during the Edo period[(4-1)]. Most of the present settlements originated from those villages. In 1889, the national government directed the adoption of a new system of town and village administrative boundaries, which combined the villages into five municipalities and later into three in 1940. In 1959, all the villages were grouped into Tadami Town, although the town is divided into three districts based on the previous delineation of 1940 village boundaries. The three districts are Tadami, Asahi and Meiwa (Figure 4.1). We conducted our field studies mainly in two areas: one along the Kurotani River in Asahi and another along the Fuzawa River in Meiwa. These two rivers are tributaries of the Ina River.

Tadami district had sustained the largest number of residents since 1959 when the current town system started. The national government initiated two dam projects for hydro-electricity in this district. Construction of Tagokura Dam, the largest dam at that time in the country, began in 1953. It was completed in 1959, while construction of Taki Dam lasted from 1959 to 1961. During those years, Tadami's population peaked at 13,106 in 1955 and decreased to 9661 in 1965. The town lost the demand for labour, and the population decrease lasted for decades until recently when it attempted to attract people from outside the town by preparing a dam tour. Table 4.1 shows that agriculture consisted of about 60% of the labour force in 1955 and declined gradually for several decades to 13.5% in 2015. In contrast, the construction sector had not declined that much in terms of the ratio. The decrease in the total number of workers was not only due to the town population decrease but also due to ageing.

About 1500 people were employed in construction in 1955, and it was over 900 in 1995. The number of people primarily engaged in agriculture decreased from about 4000 to about 600 during the same period. Still, many people grow vegetables and rice for household consumption. The relatively large share of employment in the construction sector is typically observed in most of rural

Table 4.1 Change in the percentage distribution of the labour force by industry and the total number of workers in Tadami Town

Year	*Agriculture*	*Forestry*	*Construction*	*Manufacturing*	*Tertiary industry*	*Total Workers*
1955	59.8	2.1	22.9	1.1	13.5	6695
1960	49.5	1.1	23.4	1.0	23.5	6619
1965	51.8	2.1	17.2	2.0	24.7	5022
1970	41.7	1.3	26.1	3.0	25.7	4841
1975	32.9	2.4	26.7	5.8	30.4	4155
1980	30.6	1.6	26.6	9.1	31.4	4144
1985	24.2	1.2	27.6	13.2	32.7	3924
1990	19.4	1.0	25.2	16.0	37.6	3451
1995	18.5	1.3	28.4	12.7	38.6	3294
2000	15.3	0.5	29.0	11.7	43.1	2935
2005	20.1	0.5	21.2	13.5	44.1	2712
2010	14.6	1.3	16.8	16.1	50.6	2315
2015	13.5	1.6	16.3	15.3	52.5	2467

Source: Tadami-machi-yakuba Sōmuka (2015), 2015 Population Census of Japan.

society, including remote islands (Chapter 2) since the national government provided subsidies to the relevant sections of local governments to prevent population declines as much as possible. This system has sustained not only construction companies but also politicians who obtained election funds from the companies, NGOs, and the administrative powers of relevant national and local government sections. Yet this kind of mutually beneficial system is more challenging to work these days than before, because the demand for social security has increased due to rapidly ageing Japanese population and because the national government has vast budget deficits (Inoue, 2019). It appears that such a trend emerged in the declining ratio of the construction sector (Table 4.1). In the meantime, the number of people employed in the manufacturing industry increased almost fivefold from 1955–1960 to 2010–2015 in Tadami.

Forests cover 76% of the land, including steep treeless slopes (Fukushima-ken Nōrinsuisan-bu, 2018). Treeless slopes are maintained by avalanches, scatter on mountain slopes, and are classified as forests according to the statistics. Despite such a large area of forests, the number of people working in the forestry sector is small. There are a few notable features about the forests in Tadami. First, natural forests predominate, consisting of 74.8% of the woods in 2015, while the national average was about 60%. Second, mountain slopes are generally steep and heavy snow causes many avalanches, thus the percentage of the land classified as forest but treeless is relatively large, which was 20.5% in 2015. Third, most of the forests are located far from residential settlements, in okuyama (remote mountains), and these are national forests. Forest resources had been exploited initially near settlements, before the Forestry Agency built logging roads extending far into the okuyama areas. Okuyama has larger areas of the national forests than

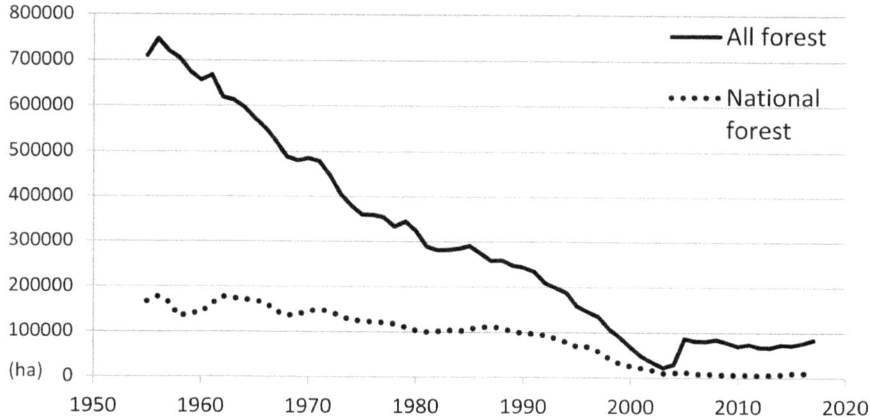

Figure 4.2 Change in the area of all forests and national forests logged in Japan. All forests include national, prefectural, municipal and private forests.

Source: Ringyō Tōkei Yōan[4-2]

does satoyama. Figure 4.2 suggests that logged area in national forests did not decline significantly until the early 1990s owing to such exploitation in okuyama and that logging of the other forests declined sharply because a substantial proportion of them was in satoyama. Nature protection groups protested against clear-cutting nearly untouched forests in okuyama across the country in the 1970s. We tried to locate old forests in our study areas in Tadami using forestry records to compare ecosystem properties between old and young forests. Still, we were able to find only a single stand that had not been cut at least for several decades in Meiwa and Asahi districts. The stand did not have many large trees because the soil layer was generally not deep enough for them to grow well (Photo 4.2).

Circumstances for the local forestry had been a bit complicated till the early 2000s when the local office of the national forest decided not to cut broadleaf trees any more for the sake of protecting the natural environment. As mentioned above, a substantial increase in demand for wood from okuyama occurred in the 1970s. The main incentive was a national government strategy to meet domestic demand for wood as much as possible (Chapter 1). Most of Tadami's broadleaf forests are not suitable for conversion to plantations, mainly due to deep snow, although the plantations have more than doubled between 1960 and 2005 (Table 4.2). Therefore, trees were cut mainly for pulp and chip wood, leaving the stands to naturally regenerate. The valley along the tributaries of the Ina and Tadami River had long logging roads except for the slopes of Mt. Asakusa[4-3]. Foresters cleared natural forests, using overhead cables to transport logs along these roads. Penna (2002) indicated that Japan's native broadleaf forests produced hardwood before they encountered a significant decline in production during the 1990s. Demand for pulp and chip wood increased during the 1970s

Photo 4.2 Beech forest nearly intact but absent of large trees.

and 1980s before reaching the maximum during the 1990s (Figure 4.3). As the area logged kept decreasing until it reached the minimum during the 2000s (Figure 4.2), probably the last remaining old natural forests were logged during these periods. Amami Oshima has a disadvantage in transportation costs relative to Honshu, and all the chip mills were closed down during the 1990s (Chapter 3). By contrast the national forests were still cut in Tadami. Because mills continued to operate in Honshu.

Table 4.2 Percentage of coniferous and broadleaf forests and areas of tree plantations, natural forests and treeless land (hectare) in Tadami

Year	*Coniferous forests (%)*	*Broadleaf forests (%)*	*Tree plantation*	*Natural forests*	*Treeless sites*
1960	2.4	97.2	1311	61,563	238
1970	3.2	96.4	1742	54,346	224
1980	5.0	88.7	2696	51,442	3639
1990	6.0	88.0	3220	51,018	3476
2000	7.0	93.0	3433	52,062	–
2005	7.0	93.0	3433	52,062	–
2012	5.1	95.3	3260	52,528	–

Treeless sites include natural grasslands and logged areas before forest recovery.

–: no statistics are available.

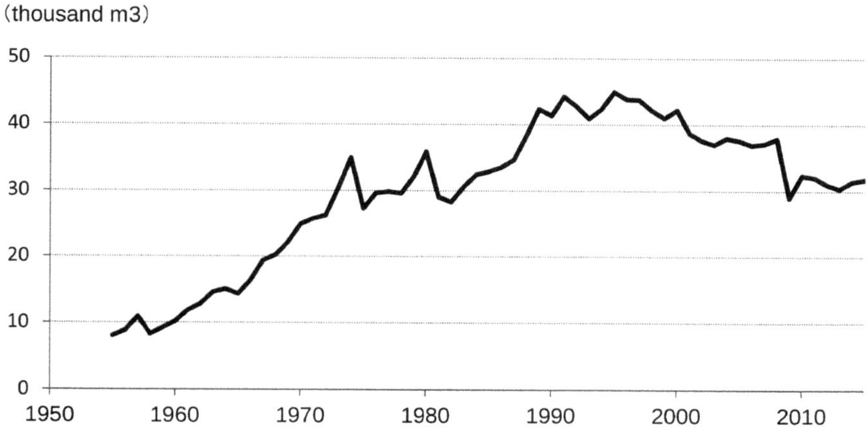

Figure 4.3 The amount of pulp and chip wood sold on the market in Japan.

Source: Ringyō Tōkei Yōan(4-2)

A local group of the Wild Bird Society of Japan found that the Forestry Agency cut an old forest stand without having permission from the prefecture. The Japanese Forest Act specifies that any activity that clears a forest stand larger than one hectare needs to have permission from the prefecture, so it was an illegal practice. The group sued the case in 2003, which local news media reported immediately. In response, the local office of the national forest decided not to cut natural forests for wood production. We were not able to find sizable old stands, even where the forest record indicated they were old, because it used to be a common practice that a forestry officer let loggers cut stands larger than the area under the logging contract. The officer received a part of the extra profits that the loggers obtained. They were not conducting such an illegal practice when the conservation group found one of these sites. However, the national forest decided to turn the past strategy, obtaining the maximum amount of production, in the opposite direction, i.e., no cutting excluding plantation trees. Receiving appeals from local conservation groups and mass media, a nature conservation NGO in Tokyo proposed an idea to set up the largest forest ecosystem protection area in the national forests in Japan (Yokoyama, 2006). Then the Forestry Agency established a committee to discuss the protection area, which was established in 2007, covering 83,890 hectares in Tadami Town and Hinoemata Village. About 70% of the area was allocated as utilisation and conservation zone, where local people were allowed to harvest wild plants and mushrooms, fish and hunt wildlife as well as to sell fishing licences to visitors. In the mid-2000s, the Tadami Town office determined a strategy for revitalising the economy through beech forest conservation and utilisation, and it succeeded in having the whole town area designated a biosphere reserve by UNESCO in 2014.

The last point which I would like to mention is the inconvenience of travel from urban areas which could be disadvantageous for the local economy. Although Japan has one of the best railway systems in the world, it takes about 4 h from Tokyo by train and about the same time by car. This feeling of the remoteness of Tadami is one of the characteristics along with mountainous landscapes surrounding the town, heavy snowfall and beech forests that encouraged young people to move out of their hometown while attracting a small number of people to migrate in.

4.1.3 *Beech forest ecosystems and wildlife species of Tadami*

The distinction between satoyama and okuyama is not very clear in Tadami. No coppice forest management is practised currently and extensive clear-cutting for pulpwood erased evidence of the past forestry practice for charcoal production. As forests cover 76% of the land and a great majority of the residential settlements are along the two rivers, most of the woods are not in typical satoyama landscapes. In other words, most of the forests are too far away to practice traditional satoyama management. Therefore, most wildlife species live primarily in typical okuyama landscapes. Here, I focus on mammal and bird species, referring to two sources of information, Tadami-chōshi Hensan-iinkai (2001) and Tadami-machi Kyōiku-iinkai (2003). I discuss their uniqueness and conservation values, taking into habitat requirements, the Red List categories they belong to, use values and observed locations of some species, occasionally in comparison with those in Amami Oshima and briefly with satoyama.

Japanese beech *Fagus crenata* is usually a single dominant canopy species in the forests of Tadami, occasionally mixed with *Quercus crispula*. The canopy height is around 20 m in old forests, while there are maples *Acer* spp. and other species of small trees along with shrubs such as *Lindera umbellate* (Family Lauraceae), *Viburnum furcatum* (Family Caprifoliaceae), ferns, bamboo grasses (*Sasa kurilensis* and *S. palmate*), and sedges growing in lower layers. The bamboo grasses often cover densely on a gentle slope, which is one of the characteristics of vegetation on the "Japan Sea" side, in contrast to the "Pacific Ocean" side. The most apparent difference between Tadami and Amami is that deciduous trees are predominant in Tadami and evergreen trees are in Amami. Due to such a difference, old Tadami forests have more complex vertical layers along with shrub and herbaceous vegetation densely occupying the understory. This difference is caused by the effects of the light environment under thin leaves of deciduous trees on the development of shrub and understory vegetation. Another feature in Tadami is that the canopy species composition varies according to terrain conditions such as a kind of walnut *Pterocarya rhoifolia*, *Aesculus turbinate* and *Cercidiphyllum japonicum*[4-4] dominant nearby streams, conifers *Pinus parviflora* and *P. densiflora* on ridges and beech on slopes. Such a variation was not very clear in Amami. In addition to these floral characteristics, mountains in Tadami have many steep slopes without tall trees but shrubs due to heavy snowfall and avalanches (Photo 4.3). In contrast, ridges and streams

Photo 4.3 Mountain slopes without tall trees but shrubs due to heavy snow and avalanches.

normally do not have steep slopes on both sides in Amami, which are unlikely to form such distinct variation as observed in Tadami.

Table 4.3 contains a list of the mammal species found in Tadami. Species richness is higher than the other regions introduced in this book, Amami, Tsukuba (northeast of Tokyo) and northern Kyushu. Many of them are endemic to Japan, and they account for about 70% of the endemics living in the main islands of Japan, i.e., excluding the Nansei Archipelago and Ogasawara Islands. In another calculation, the mammals in Tadami and Amami combined account for 60% of the endemic species inhabiting Japan. It may be surprising that such small areas contain such a high proportion of endemic species. The table includes all the mammals larger than rabbit living in Japan, except for brown bear in Hokkaido, i.e., bear, deer, serow, monkey, fox, raccoon dog and badger. Wild boar is not in the table, but they are found these days due to decreasing winter snow depth. Biogeographical isolation has not allowed brown bears to move to Honshu (Chapter 1). Six arboreal species require cavities for nesting; *Sciurus lis*, *Petaurista leucogenys*, *Pteromys mononga*, *Glirulus japonicas*, *Myotis pruinosus* and *Murina ussuriensis*. Their distribution ranges are large, and none of them is endangered in contrast to long-haired rat *Diplothrix legata* in the Nansei Archipelago, which is the only arboreal mammal species there. World Census of Agriculture and Forestry reported that natural forests over 61 years old ranged from 36 to 39 thousand hectares in Tadami and they accounted for 68, 66 and 71% of all the forests in 1980, 1990 and 2000, respectively. Although the forest records are not reliable as I indicated before, we found some old trees in some stands that had been illegally logged. Because forested areas are so large, there would be many large trees left uncut away from logging roads. Heavy snow

Table 4.3 Mammal species reported to live in Tadami

Common name[a]	*Scientific name*	*Approximate locations of distribution*[b] *Categories in the Japanese Red List*
Shrew	*Chimarrogale platycephala*	Honshu, Kyushu
	Crocidura dsinezumi	Hokkaido to Kyushu
Mole	*Dymecodon pilirostris*	Mountains in Honshu, Shikoku, Kyushu
	Urotrichus talpoides	Honshu to an island south of Kyushu
	Mogera imaizumii	Mountains in Honshu, Shikoku
Bat	*Rhinolophus cornutus*	Hokkaido to Kyushu, Amami Islands
	Rhinolophus ferrumequinum	(Eurasia)[c]
	Pipistrellus endoi	Honshu, Shikoku :VU
	Myotis pruinosus	Honshu, Shikoku, Kyushu :VU
	Myotis macrodactylus	(China, Korea, Japan)
	Murina ussuriensis	(China, Korea, Japan)
Mouse	*Eothenomys andersoni*	Mountains in Honshu
	Eothenomys smithii	Honshu, Shikoku, Kyushu
	Apodemus speciosus	Hokkaido to Kyushu
	Apodemus argenteus	Hokkaido to Kyushu
	Mus musculus	(Worldwide)
Rat	*Rattus norvegicus*	(Worldwide)
Squirrel	*Sciurus lis*	Honshu, Shikoku
Flying squirrel	*Petaurista leucogenys*	Honshu, Shikoku, Kyushu
	Pteromys mononga	Honshu, Shikoku, Kyushu
Dormouse	*Glirulus japonicus*	Honshu, Shikoku, Kyushu
Hare	*Lepus brachyurus*	Honshu, Shikoku, Kyushu
Monkey	*Macaca fuscata*	Honshu to an island south of Kyushu
Bear	*Ursus thibetanus*	(East Eurasia)
Fox	*Vulpes vulpes*	(Holarctic)
Raccoon dog	*Nyctereutes procyonoides*	(East Asia)
Weasel	*Mustela itatsi*	Honshu, Shikoku, Kyushu
Stoat	*Mustela erminea*	(Holarctic)
	Martes melampus	Honshu, Shikoku, Kyushu
Badger	*Meles anakuma*	Honshu, Shikoku, Kyushu
Civet	*Paguma larvata*	(Introduced from Asia)
Deer	*Cervus nippon*	(East Asia)
Serow	*Capricornis crispus*	Honshu, Shikoku, Kyushu

Abbreviation: VU, vulnerable.

Notes:

a Common family or group name more general than species name.
b Introduced area is not included.
c Information in parentheses were retrieved from the IUCN Red List of threatened species, https://www.iucnredlist.org, 12/12/2019.

There are two species that may have been alive but extinct. An interview survey suggested that river otter *Lutra lutra* lived until the 1950s. There is no clear evidence of wolves *Canis lupus*, which crossed with dogs and Tadami's three records do not reliably identify the species.

Source: Tadami-chōshi Hensan-iinkai (2001); Tadami-machi Kyōiku-iinkai (2003); Abe (2008); Kankyō-shō (2018).

prevented natural forests from being converted to extensive tree plantations which do not provide good habitat conditions for mammals in terms of the food supply. There are no documents on the past forestry operations, and we were able to visit only a small portion of the whole forest area in Tadami. Many old broadleaf trees left uncut must have provided sufficient food and nesting sites to maintain rich mammal species diversity. Such situations contrast with Amami, where the area of old forests was much smaller in the early 1990s. During this period, only the international economy and a private company's strategy contributed to protecting the remaining populations of arboreal species.

Tadami-chōshi Hensan-iinkai (2001) reported 118 bird species, excluding transients and those recorded no more than a few times. They belong to 40 families (Table 4.4) and account for more than half the number of species listed in

Table 4.4 Avian families and the number of species in each family observed in Tadami excluding transients and those recorded no more than a few times

Family name	*Common name*[a]	*Number of species*[b]	*Red List, if any*[c]	*Habitat*[d]
Accipitridae	Hawk and eagle	11 (3N, 7S, W)	2EN, 2VU, 4NT	Forest, rural
Falconidae	Falcon	1 (N)	VU	Forest
Phasianidae	Pheasant	1 (N)		Forest
Columbidae	Pigeon	2 (S)		Forest, rural
Cuculidae	Cuckoo	4 (S)		Forest
Strigidae	Owl	2 (N, S)		Forest
Caprimulgidae	Nightjar	1 (S)		Forest
Apodidae	Swift	2 (S)		Mountain
Alcedinidae	Kingfisher	3 (2N, S)		Forest
Coraciidae	Roller	1 (S)	EN	Forest
Picidae	Woodpecker	4 (N)		Forest
Hirndinidae	Swallow	2 (S)		Rural, mountain
Motacillidae	Wagtail	4 (N, 3S)		Rural, riverside
Campephagidae	Minivet	1 (S)	VU	Forest
Pycnonotidae	Bulbul	1 (S)		Forest, rural
Laniidae	Shrike	2 (S)	CR	Forest, rural
Cinclidae	Dipper	1 (N)		Riverside
Troglodytidae	Wren	1 (S)		Riverside
Turdidae	Thrush	8 (6S, 2W)		Forest
Sylviidae	Warbler	7 (N, 6S)		Forest, wetland
Muscicapidae	Old World flycatcher	4 (S)		Forest
Aegithalidae	Long-tailed tit	1 (N)		Forest
Paridae	Tit	4 (N)		Forest
Sittidae	Nuthatch	1 (N)		Forest
Certhiidae	Tree creeper	1 (N)		Forest
Zosteropidae	White-eye	1 (S)		Forest, rural
Emberizidae	Bunting	5 (4S, W)	NT	Forest
Fringillidae	Finch	4 (N, 2S, W)		Forest

(*Continued*)

Table 4.4 Avian families and the number of species in each family observed in Tadami excluding transients and those recorded no more than a few times (*Continued*)

Family name	*Common name*[a]	*Number of species*[b]	*Red List, if any*[c]	*Habitat*[d]
Ploceidae	Sparrow	2 (N,S)		Forest, rural
Alaudidae	Sky lark	1 (S)		Rural
Sturnidae	Starling	2 (S)		Rural
Corvidae	Crow	4 (3N, S)		Forest, rural
Ardeidae	Heron	5 (2N, 3S)		Rice paddy
Rallidae	Rail	1 (S)		Wetland
Charadriidae	Plover	1 (S)		Wetland
Scolopacidae	Sandpiper	1 (S)		Wetland
Podicipedidae	Grebe	1 (N)		Lake
Phalacrocoracidae	Cormorant	1 (N)		Lake
Anatidae	Duck	16 (2N, 14W)	VU	Lake, rice paddy
Laridae	Gull	3 (N, 2W)		Lake

Notes:

a If the family has more than one common name in English, the one that matches the Japanese species is adopted.

b N: non-migratory, S: spring, W: winter.

c CR: critically endangered, EN: endangered, VU: vulnerable, NT: near threatened according to the Japanese Red List 2019.

d Main habitat types the species require; "rural" covers farmland, residential areas and forest edges; "mountain" indicates steep slopes without trees.

The CR species is *Lanius tigrinus*; three species of EN are *Spizaetus nipalensis*, *Aquila chrysaetos* and *Eurystomus orientalis*; four species of VU are *Anas formosa*, *Haliaeetus albicilla*, *Butastur indicus* and *Pericrocotus divaricatus*); five NT species are *Pandion haliaetus*, *Pernis apivorus*, *Accipiter gentilis*, *Accipiter gularis* and *Emberiza sulphurata*.

the Japanese wildlife inventory (Kankyo-shō, 2002). One of the attributes that imply richness of the avifauna is eleven species of Accipitridae, most of which live in forests and treeless mountain areas. There are 16 species of Accipitridae in the Japanese wildlife inventory, excluding transients and the migrants that seldom fly to Japan. They occupy the highest trophic levels, suggesting richness of prey species living in Tadami. One reason why Tadami has such diversity of raptors is that there are many treeless mountain slopes developed by avalanches due to heavy snowfall. The mixture of treeless slopes and forests would provide a variety of habitats, resulting in high species richness of prey and predators. Besides, natural forests are predominant over single-species even-aged conifer plantations. Natural forests of deciduous trees not only provide more food than plantations, such as berries, nuts and acorns but also form the more complicated vertical structure of the woods. Seasonal variation in bird species composition is considerable. The table contains 34 non-migrant, 63 summer and 21 winter birds. The majority of the winter birds live in three dam lakes, and there is no natural lake or pond in Tadami.

It is not very easy to clarify the conservation value of the fauna and flora living in the forests of Tadami to protect them from forest resource exploitation and other development activities because there are few species or populations unique to Tadami, as is the case on Amami Oshima. Ninan and Inoue (2013) attached no value to biodiversity when they evaluated the ecosystem services of the protected areas of the national forests in Tadami. The town is the most interested in preserving a kind of willow tree, *Salix hukaoana*. It lives along river banks located outside forested areas, so it cannot contribute to forest conservation. My opinion is that mammal fauna, collectively, can be a good candidate if a talented movie editor makes an attractive film that illustrates its uniqueness as an entity to draw interest from conservationists and nature lovers living outside the town. Amami rabbit has enticed many scientists, conservationists and tourists from outside, although spiny rat should obtain higher evaluation than the rabbit if evaluated objectively based on degree of rareness and taxonomic uniqueness (Chapter 2). On the other hand, birds played the most important role in generating support for protecting the natural forests of Tadami, since it was a small local group of ornithologists who initiated the action to stop clear-cutting of the national forests. I do not find a good academic reason to evaluate the birds higher than the other taxonomic groups. Also, fortunately, it was the time when the Forestry Agency changed its fundamental policy from timber-oriented forestry to such an approach that treats every forest function equally as a management objective.

4.2 Biodiversity-based ecosystem services before the great earthquake

In terms of forest ecosystem services affecting local people's lives, the most visible one is harvesting wild plants and mushrooms. A previous study on this subject (Ninan and Inoue, 2013) maintained that water-related services such as flood control and water supply were far greater than forest recreation. The study's evaluation of ecosystem services was based on the construction cost of the Tagokura Dam, which was built primarily for producing electricity. They supposed the cost was equal to the value of ecosystem services and enlarged it to a massive amount of value using the price difference between the period of construction and that of their study. Therefore, my observation and the conclusion of the previous research are very different from each other. There was an incident that contradicted their evaluation when the town suffered from record rainfall, totalling 711.5 mm in four days of late July in 2011. Discharged water from Tagokura Dam flooded the town area of Tadami and destroyed several bridges downstream. One of them was a railway bridge, and the destruction has stopped the train for nine years since. Besides, as for water supply, most town people use other watershed areas for drinking and agricultural irrigation. Thus, it is not reasonable to account for the value of forest ecosystem services in terms of the cost of dam construction.

Local people in Tohoku region go into forests much more often than those in Kyushu. After the snow melts in spring, wild plants leaf out on a treeless slope and along a forest road. Their juvenile leaves are soft, not very bitter and good for salad, tempura, salted food, etc. New leaves and buds of herbs, ferns and trees are available, as long as they know how to process and cook, which is passed on from generation to generation as traditional knowledge. Tempura, deep-frying, is the most popular way of cooking wild plants, while they usually cook mushrooms for soup with soy sauce or soybean paste. The main harvesting period is from early May to the middle of June, starting from a lower elevation, going up higher as the snow melts gradually and changing species harvested as well. June and July are the peak months for angling. Hot weather and horseflies prevent people from staying in forests except for mountain hikers passing through. Then they start looking for mushrooms in September, which last till November. Forests provide adequate conditions for mushroom picking as there are many decaying tree trunks and a variety of broadleaf trees that have symbiotic relationships with mushrooms. Snow season is good for hunting and watching wildlife. This kind of lifestyle is called "following the Jomon calendar." Jomon is the period before rice agriculture was introduced to Japan when population density was higher in Tohoku than in Kyushu, from about 5000 BC to 2000 BC (Koyama and Sugito, 1984).

4.2.1 *Harvesting wild plants and mushrooms*

From 2008 to 2011, right before the great earthquake, I had a research project, whose main objectives were to develop methodologies to evaluate ecosystem services quantitatively and to compare the services between natural forests and tree plantations. To make quantitative evaluations, we conducted questionnaire surveys, asking local people to carry a small handheld Global Positioning System (GPS) device and report how much non-timber forest products (NTFP) they gathered. I introduce below some of the results reported in Matsuura *et al.* (2013). We selected two areas in Tadami, one located along the Kurotani River in Asahi and the other along the Fuzawa River in Meiwa (Figure 4.1). The questionnaire study divided the respondents into six groups according to sex and age, i.e., under 60, from 60 to 69 and no younger than 80. The results indicated that 50–90% of the households had at least one family member, who went out or used to go frequently to harvest at least either wild plants in spring or mushrooms in fall. Those who went out most often were the males of the second age group, i.e., from 60 to 69. Many of them would have retired from full-time works and were still physically strong enough to go up and down mountain slopes. In the mountains of Tohoku and Hokkaido, more than a few people are killed each year and many are injured by bears. Males went out more often than females, probably because females are busier at home, and they prefer to be accompanied by her male partner as a guardian.

In spring after snowmelt, local people start harvesting wild vegetables, using forest roads reached higher than 1000 m above sea level. The local research participants collected 16 favourite species of plants (Table 4.5). Half of them do not have close relatives in Europe or North America, and I did not find any suitable

Table 4.5 Edible wild plants harvested in spring and summer in Tadami arranged in descending order of popularity[a]

Scientific name	*Japanese name*	*English name*	*Edible part*
Matteuccia struthiopteris	kusasotetsu	Ostrich fern	Young shoot
Aralia cordata	udo	Japanese-asparagus	Young shoot
Pteridium aquilinum	warabi	Bracken[b]	Young shoot
Petasites japonicus	fuki	Giant butterbur	Bud, stalk
Osmunda japonica	zenmai	Royal fern[b]	Young shoot
Hosta montana	oba-giboushi	None	Leaf and stalk
Eleutherococcus sciadophylloides	koshiabura	None	Young shoot
Aralia elata	taranoki	Angelica-tree	Young shoot
Laportea macrostachya	miyama-irakusa	None	Leaf and stem
Elatostema umbellatum	uwabamisou	None	Stalk
Parasenecio delphiniifolius	momijigasa	None	Leaf and stem
Smilax riparia var.ussuriensis	shiode	None	Young shoot
Sasa kurilensis	Chishima-zasa	None	Young stalk
Parasenecio hastatus	inudouna	None	Young shoot
Osmundastrum cinnamomea	yamadori-zenmai	Buckhorn	Young shoot
Allium victorialis	gyouja-nin'niku	Alpine leek	Leaf and stalk

a Popularity is determined according to the number of local pickers reported by Matsuura *et al.* (2013).

b English names are referred to Global Biodiversity Information Facility[(4-5)] and Matsuura *et al.* (2014).

words for translation. They harvested most of the plants when they were young and ignored them afterwards. Local people knew that each species had specific habitat preference as well as the best time for harvesting. For instance, they looked for *Osmunda japonica* and *Hosta montana* on a slope, where avalanches had prohibited trees growing, *Aralia cordata* along a forest edge on a road or in treefall gaps, and *Pteridium aquilinum* in grassland. The duration of harvesting depended on the range of altitude in which a species grew, and the maximum altitude harvesters could go up. They parked their car on a roadside and walked along a small stream, where edible plants were most numerous, usually for less than one kilometre because moist plants were heavy. We also inspected whether harvesting has significant effects on the abundance of four popular species of plants by comparing appearance frequency (probability of appearance at a given spot) between areas near forest roads and those farther from roads. The analysis found no statistical difference between the two areas (T. Matsuura, personal comm.); in other words, no significant impacts of harvesting on the distribution range of those species. Local people sometimes avoid overharvesting intentionally. For instance, they do not cut off all the shoots of *Matteuccia struthiopteris* but leaves some in the middle. Oppositely, *Petasites japonicus* and *Pteridium aquilinum* are too numerous to worry about overharvesting. In the past, they could sell some species at a much higher price than now.

Source of income ranked low among the reasons for going out, compared to enjoying a sense of seasonal change, obtaining healthy food, hiking, the delight of searching for plants, making other people happy by sharing with neighbours or friends living in urban areas (Matsuura *et al.*, 2013). There used to be rich people in Tadami district, who had made a lot of money by gathering and selling *O. japonica*. Some of them lived in a cottage in mountains with steep slope valleys far from town for a month or two. Pre-sale treatments take a lot of time and efforts, but it was worthwhile. Many other species were also precious as the locals were much more dependent on various forest resources and rules prohibiting harvesting them before a certain date determined by the settlement community (K. Kariya, personal comm.). The date varied from year to year according to the abundance estimated by the community.

In contrast with wild vegetables, most research participants disagreed with carrying GPS devices and providing information on the spots where they picked mushrooms. That was because they had their treasure maps in their memory and wanted to monopolise mushrooms on the spots that only they knew and visited each year. Mushroom hunting is unpredictable and opportunistic, but some trees are more likely to grow a certain kind of mushroom. If they find a strain, which may be too early to pick up, they try to hide it, leaving no footprints, but often in vain. Matsuura *et al.* (2013) reported that the best season was from late September to the end of October. The local participants collected 19 favourite species (Table 4.6).

There must be many more edible species, as Yamada (2002) found in a literature review that there were more than 300 wild species of potentially edible symbiotic fungi in Japan. I understand that Tadami has extravagantly rich fungi resources, so the locals ignore many kinds of edible fungi they do not prefer to eat. The knowledge transferred from generation to generation must have produced significant differences in the species frequently harvested from a local community to another. Table 4.6 shows that most species depend on broadleaf forests. Many of them grow on decaying trees, and others are symbiotic living trees. Unlike the southwestern part of Japan, where conifer plantations occupy large areas of okuyama, forests in Tadami are able to maintain habitats for many kinds of edible fungi. After the introduction of rice farming, rural regions in Tohoku were generally less populated than in the southwest, due to the cold climate and long lasting snow cover. High population density in the southwest demanded a large amount of fuelwood from coppiced forests, which did not have many decaying trees. *Tricholoma matsutake* is symbiotic with pines and is sold at a very high price, often over US $500/kg. Pines grow along ridges, where the soil has a shallow surface layer of raw humus. Matsutake is found on the roots of *Pinus densiflora* and less frequently under *Pinus parviflora*. It is so expensive that there are matsutake pickers as if it is a temporal profession. They go out so early in the morning that the early bird may catch the worm. One local community decided to make money by collecting a fee from outsiders entering the forests they manage. Then groups of 10–20 people came, lined up in a row, slowly climbed upslope and picked all the matsutake found. The community found the resource was depleted within a couple of years, and finally, they decided to close the forests to await the recovery of the matsutake resource.

Table 4.6 Edible mushrooms harvested in Tadami arranged in descending order of popularity

Scientific name	*Japanese name*	*English name*	*Habitat preference*
Pholiota nameko	nameko	Scalycap)	Decaying beech tree
Mycoleptodonoides aitchisonii	buna-haritake	None)	Decaying beech tree
Grifola frondosa	maitake	Hen of the woods	Base or above ground root of a large beech family tree
Armillaria spp.	naratake	Honey fungus	Decaying broad-leafed or conifer tree
Sarcomixa serotina	mukitake	Olive oysterling	Decaying broad-leafed tree
Pleurotus ostreatus	hiratake	Oyster mushroom	Decaying broad-leafed tree
Lyophyllum decastes	hatake-shimeji	Clustered domecap	Forest floor, grassland
Tricholoma matsutake	matsutake	Spicy knight	Pine forest floor
Sarcodon aspratus	koutake	Violet tooth, etc.)	Mixed broad-leafed-pine forest floor
Hypholoma lateritium	kuritake	Brick tuft	Decaying broad-leafed tree
Laetiporus sulphurous	masutake	Chicken of the woods	Broad-leafed living or decaying tree
Lyophyllum fumosum	shaka-shimeji	Domecap	Mixed beech family-pine forest floor
Meripilus giganteus	tombi-maitake	Giant polypore	Base or above ground root of a large beech tree
Auricularia auricula-judae	kikurage	Jelly ear	Broad-leafed tree
Climacodon septentrionalis	ezo-haritake	Northern tooth fungus	Beech or maple tree trunk
Lyophyllum shimeji	honn-shimeji	Domecap)	*Quercus serrata* or mixed *Q. serrata- Pinus densiflora* forest floor
Lactarius volemus	chichitake	Fishy milkcap	Broad-leafed forest floor
Hypsizygus marmoreus	buna-shimeji	Beech leech	Decaying broad-leafed tree
Cortinarius tenuipes	kuri-fuusentake	Webcap)	Oak forest floor

Reference: Matsuura *et al.* (2013) for the species harvested.

English names are referred to British Mycological Society[4-6]:

An English name given in half parenthesis indicates that there is no name for the species but one for the same genus species. If described as "none," the genus is found neither in Europe nor in the North America.

Imazeki *et al.* (2011) is referred to for the habitat preference.

In addition, we conducted in 2008 a questionnaire survey in a village next to Tadami, Hinoemata, in order to evaluate the economic contribution of NTFP to the local community. The village is 700 m above the sea, 400 m higher than Tadami, and the area is 390 km^2, in which a national park occupies a little less than half of the village land (Photo 4.4). Forest ecosystems are similar to those in Tadami, but with long winters, less snow and smaller areas of tree-less slopes. Urban travellers enjoy eating wild plants, mushrooms, stream fish,

Photo 4.4 Hinoemata Village surrounded by Oze National Park.

wildlife meats and even salamander tempura with curiosity in addition to its natural beauty and hot springs. In the survey, 90% of 110 households answered that one of their family members went to forests for at least one of the following activities on average of 20 days a year: wild plant and mushroom harvesting, fishing and hunting. Running a guest house makes the greatest contribution to the village economy, and the survey found that wild foods gathered in the forest accounted for more than half of the dishes served to guests. In the same year, 32,000 people stayed overnight in the village with 32 guest houses and about 650 inhabitants. Therefore, NTFPs must have made an outstanding contribution to this small village.

4.2.2 *Nature-based recreation and tourism in Fukushima Prefecture*

From fall 2007 to spring 2010, I drove a total of about 500 km through forested areas in Tadami and surrounding towns and villages, counting cars parked along the roads and visitors enjoying recreational activities and scenic beauty. I also drove about 170 km in forests in the southeastern part of the Abukuma Mountains. The surveys were very time-consuming, and I was unable to control conditions such as weather, repetitions per route, time of day for visits. The only requirement was narrow routes with no central line and forests on each side. Observations along the roads revealed that visitors were involved in a variety of activities, such as harvesting wild plants, picking mushrooms, angling, hiking, catching insects and driving a car to enjoy the scenery (Table 4.7). Despite such

Table 4.7 Types and frequency of activities observed in the southern Aizu region and Abukuma Mountains (per 100 km)

	Frequency	
Activity	*Aizu*[a]	*Abukuma*[a]
NTFP harvesting	10.4	2.4
Fishing	8.0	3.5
Hiking[b]	9.0	0.6
Driving	11.0	4.1
Insect collecting	0.6	1.2

Notes:

a The survey routes were in six municipalities in Aizu and four in Abukuma (see Figure 4.1).

b Most hikes are day trips to the summit.

a crude method, the comparison with Abukuma gives some insights on the components of natural environments that visitors preferred in the southwestern part of Fukushima. There were many more visitors in southern Aizu than in the Abukuma study areas, although the six municipalities in Aizu had a population of 29,000, which was much smaller than four cities in Abukuma, 477,000, as of 2010[(4-7)]. In Aizu, most visitors were from the Tokyo metropolitan area, whereas Abukuma had fewer visitors from the Tokyo area. The critical factors of the forest ecosystems that made such a large difference must have been snow depth, mountain height and dominant forest types. Access from Tokyo to the Abukuma study areas is significantly better than Tadami, but with little impacts.

Stream fishing was one of the most popular activities. There are three species of fish that are particularly popular due to their abundance and taste for food: *Salvelinus leucomaenis* (char), *Oncorhynchus masou* (trout) and *Cottus pollux* (sculpin). All of them are adapted to cold water and do not swim downstream to the sea. Prefectural governments give inland water fishery cooperatives exclusive fishing rights, instead demanding resource management through breeding and releasing fry. There are six cooperatives in Fukushima Prefecture. They release two species of fish in June, char and trout, and the open fishing season ends in September. The cooperatives charge ¥1100 (ca. 10 US$) for the pre-purchased one-day license and ¥1600 for the on-site license. The total sales amount in Tadami in 2009 was ¥531,000. Stream fish feed on aquatic and terrestrial insects falling from the upper canopy. Yoshimura (2010) found that coniferous forests were richer in aquatic insects than broadleaf forests, but no significant correlation between the abundance of aquatic insects and fish population density. Therefore, there is no reason to believe that either forest type or insect abundance provides a better habitat for fish. Broadleaf tree canopy can cover a stream larger than conifers, and falling insects may provide fish with sufficient food. On the other hand, I located more anglers, mostly men, in deciduous broadleaf forests than in conifer plantations (Figure 4.4). Deciduous broadleaf forests cover

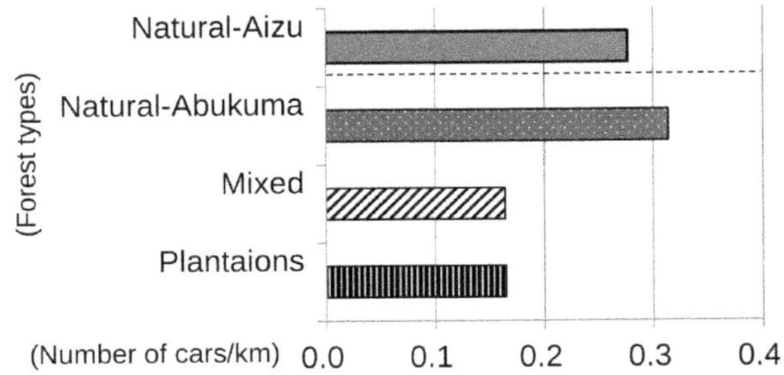

Figure 4.4 Comparison of the number of anglers between the southwestern part of Fukushima and Pacific coast areas.

much larger areas in and around Tadami than the Abukuma Mountains near the Pacific Ocean, where plantations occupied about half of the forests. Vehicle registration plates(4-8) showed that visitors to Aizu generally came from distant urban areas, and in Abukuma from nearby cities. I suppose they preferred the kind of forest atmosphere that deciduous trees created to fishing under conifers of dark green and thick leaves.

The number of cars parked near the entrance to the hiking route showed the popularity of hiking. I examined it at the foot of five mountains in Aizu (three in Tadami and two in Hinoemata) and three mountains in Abukuma. The mountains in Aizu were much more popular than those in Abukuma (Table 4.7). There are three apparent differences in environmental factors between them; height, forest type and amount of snow in early summer. As shown in Table 4.8, more visitors came to Tadami in October than in June. The statistics were taken by a museum located near a popular tourism destination with magnificent views of the mountains and the dam lake. Because there is no public transportation to the dam and museum, almost all the visitors went there by private automobiles or rental cars. Thus the statistics were thought to indicate a seasonal variation in the number of people travelling by car. The table suggests that autumn colours of deciduous leaves attract more people than fresh green leaves in May when snow remains on the slopes of high mountains.

I saw insect collectors but have never seen anybody enjoying bird watching in this study region even though the species diversity was high. In contrast, I met amateur bird watchers in Amami, even from Europe, North America and Oceania. On the other hand, a local bird-watching group had a significant impact on the protection of national forests from clear-cutting, as mentioned above.

Table 4.8 Number of visitors to Beech Museum in Tadami from April 2014 to March 2015

Month	*4*	*5*	*6*	*7*	*8*	*9*	*10*	*11*	*12*	*1*	*2*	*3*
Number of visitors	171	478	771	637	1153	649	1279	635	93	184	397	384

Courtesy of S. Kawarasaki for providing statistics.

In another case, they stopped constructing a forest road, claiming that it would disrupt the nesting habitats of the golden eagles (*Aquila chrysaetos*), which is considered so unique that the species is almost always included in the national environmental impact assessment process.

In summary, the survey results suggest that deciduous broadleaf forests provide suitable conditions for nature-based recreation, at least angling and tourism. Besides, old natural forests provide habitats for many kinds of mushrooms. Heavy snowfalls and steep slopes maintain grasslands where edible wild plants grow. These conditions are unsuitable for tree plantations and help maintain large areas of natural forests forming unique landscapes dominated by beech forests and treeless slopes. Most of these types of landscapes are remote from human settlements and provide suitable habitats for many species of wildlife, some of which are either endemic to Japan or ranked relatively high in the national Red List. Therefore, anthropogenic intervention is unnecessary for biodiversity conservation and biodiversity-based forest ecosystem services in Tadami, except to maintain forest roads for forest use.

4.3 Effects of the nuclear power plant accident and population decrease

In March 2011, Fukushima Daiichi (Number One) nuclear power plant suffered a hydrogen explosion that contaminated wide areas with radioactive Cs (Yoshida and Kanda, 2012). The contamination level depended not only on the distance from the power plant but also on the wind direction that changed over time after the explosion. A direction change resulted in high levels of contamination in Iidate Village situated northwest of the plant first, then, another wind shift contaminated Naka-douri. Thus areas north of Tokyo had relatively high levels of radioactive Cs due to rainfall occurring a few days after the explosion. Researchers in various ecological fields studied the behaviour of radioactive Cs after the accident, such as accumulation in soil layers (Ikeda *et al.*, 2014), discharge into rivers (Ochiai *et al.*, 2015), movement within a tree (Yoshihara *et al.*, 2013) and absorption by herbaceous plants (Kiyono and Akama, 2013), mammals (Yamada *et al.*, 2013), insects (Nishikiori *et al.*, 2015) and other living organisms. Kaneko *et al.* (2014) concluded that rainwater leached Cs or moved to the forest floor along with fallen leaves and twigs. They also indicated that the decomposition of organic matter brought Cs into the soil, some of which was absorbed by edible wild vegetables and mushrooms. IAEA (2006) suggested that most Cs would stay in ecosystems for an extended period, some of them moving back and forth between soil and trees. Wild boars contain high levels of radioactive Cs, probably due to feeding habits, and many locals noted increases in boar numbers since the contamination had reduced hunting pressure.

Leaving contaminated forests untouched is an unacceptable option for locals, but the problem is that the affected areas are too large to treat. The Ministry of the Environment and the Tokyo Electric Power Company have jointly invested in removing Cs-contaminated soil from residential areas and farmlands.

In contrast, forests have remained untouched unless they were within 20 m of residential areas, farmlands or main roadways. Removing all surface soil was too costly for its effectiveness. Another reason was that the company and the ministry did not realise the importance of forests for local residents. As such, there are few studies on forest ecosystem services except for wood production (e.g., Kimura, 2016), even though there are many studies on ecosystem functions. It is conventional attitudes among specialists in physical and biological sciences, who are interested in those aspects of science but not much in human society and its connection to ecosystems that studies of how changes to those ecosystems affect people are scarce. As far as I perceive, locals and visitors use forests most frequently in terms of cultural services, as shown in Table 4.7. In the following section, I compare the services before and after 2011 in Aizu and Abukuma regions. I also looked over the whole prefecture and attempted to discern the effects of the accident on cultural services. I also tracked the Cs level changes in certain species of wild plants, mushrooms, fish and wildlife. I used several information sources, such as tourism statistics, sighting forest users and official reports on the radioactive Cs.

4.3.1 *Effects on cultural services in Fukushima Prefecture*

If any NTFP has radioactive Cs exceeding 100 Bq/kg-wet (Cs-134 and 137 combined), it is prohibited not only to sell but also serve it in guest houses. Each municipality has inspection facilities to measure the radioactivity of raw or processed food samples. Generally, mushrooms have much higher levels of Cs than wild vegetables. If only one sample of a wild mushroom species exceeds the limit, the regulation applies to all species. When inspecting a large number of pieces, it is likely to have specimens exceeding the criterion. Once this control starts, town or village offices must meet the following conditions to restore commercial use of mushrooms. First, the local government has to select the species to be exempt from the restriction and set up five monitoring points for each species. Next, five samples should be collected, and each piece must weigh no less than 200 g. Third, all samples must meet the food safety standard, and the average Cs level must decline each year. The national Food Agency clears each species limit one by one. It would be hard to collect a given amount of mushrooms on the same plot since their production is uncertain. All municipalities in Hama-douri and Naka-douri, 13 and 29 municipalities, respectively, suffered the restriction by the end of 2014. In contrast, 9 of 14 municipalities suffered the restriction in Aizu. As of the end of 2019, the government removed the restriction for six species in Tadami Town. In 2013, the proportion of the mushroom samples exceeding the limit reached a peak of 74% even in Aizu. Then it gradually decreased to 26% in 2014, 10% in 2015 and 0% in 2016. Seven mushroom species had a much lower level of Cs in 2017 than the values from 2011 to 2013 (Figure 4.5). In 2017, I conducted a personal interview survey in three towns in Aizu, which indicated that enthusiastic mushroom pickers went out as often as they did before 2011.

The situations around the power plant are not comparable with those in Aizu. I conducted a small scale survey in Kawauchi Village from 2015 to 2018

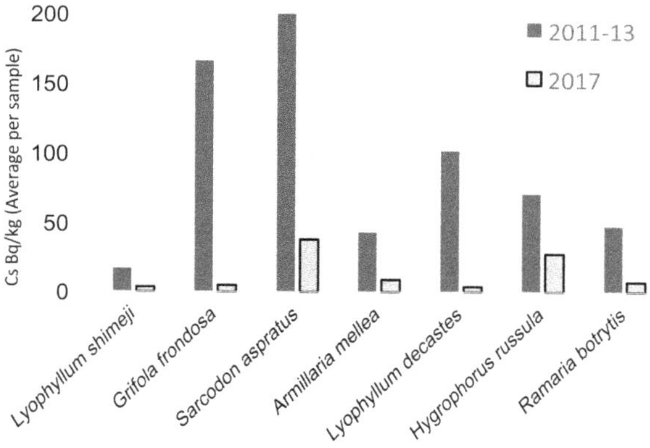

Figure 4.5 Decrease in the caesium level in some mushroom species of local favourite in Aizu.

and Tomioka Town from late 2018 to early 2019. The residential settlements in Kawauchi are about 15–25 km southwest of the power plant in small valleys upstream from the Pacific Ocean. The registered population in March 2011was about 3000, and the central government instructed all residents to evacuate from the village right after the accident. Thanks to the wind direction, the village area was not very seriously contaminated, and most of the residents were able to return in about a year. The registered population decreased to 2677 in 2018, of which 512 did not live in the village[(4-9)]. Tomioka Town is south of the power plant, no farther than 13 km. The population was about 16,000 and about three-quarters of them still registered as residents as of February 2019. Yet I understand that only about 500 had returned and lived there. Radioactive Cs levels remained high in Hama-douri in 2018 (Table 4.9).

I interviewed 12 people who had picked mushrooms frequently in Kawauchi before 2011. Four of them still went out often as of 2018, and three of them lived

Table 4.9 Mean and standard deviation of radioactive caesium (134 and 137 combined) in wild mushroom samples measured at the inspection centre in Kawauchi Village in 2017 and 2018

Scientific name	*Japanese name*	*English name*	*Radioactive Cs (Bq/kg)*
Cortinarius caperatus	shougenji	The gypsy	5195 ± 1105 (5)
Hygrophorus russula	sakura shimeji	Pinkmottle woodwax	1888 ± 1004 (12)
Sarcodon aspratus	koutake	Violet tooth, etc.	1402 ± 1196 (35)
Suillus bovinus	amitake	Bovine bolete	425 ± 509 (41)
Lyophyllum fumosum	shaka shimeji	Domecap	118 ± 82 (13)
Armillaria mellea	naratake	Honey fungus	86 ± 122 (5)
Albatrellus confluens	ningyoutake	Polypore	53 ± 52 (8)

Sample size is in parenthesis.

Refer to Table 4.6 for the explanation of the English names.

in the district farthest away from the power plant. Their favourite recipe is boiling koutake *Sarcodon aspratus* and mixing it with rice. Sixteen samples of koutake brought to the inspection centre revealed an average of 1524 Bq/kg of radioactive Cs-137. They knew such a high level but could not help eating the mushroom. One of my colleagues, Toshiya Matsuura, and I conducted a questionnaire survey in Kawauchi, and we found many comments in the optional free-text field. They exposed their feeling of despair for the lost opportunity to go to forests and enjoy wild plant and mushroom harvesting. The number of such respondents was 61 out of 355, an exceptionally large number of optional comments, suggesting that the accident brought about unexpected extreme losses of cultural forest services.

The Pacific coast region has much less snowfall than the western part of Fukushima Prefecture. The Building Standard Act gives a good example of the difference. The prefecture requires houses to withstand 350 cm of snow according to this act in Aizu, while it is 90 cm in Kawauchi Village. Such a climatic difference encouraged silvicultural activities in Hama-douri and resulted in considerably higher ratio of plantation coverage and younger natural forest stands than in Aizu. Yet such differences do not seem to affect the harvested species, and interviews with locals revealed that most of the species were favoured in both Aizu and Hama-douri. Only two species, *Matteuccia struthiopteris* and *Osmunda japonica*, were not frequently harvested in Hama-douri. Before starting this study, I assumed that mushrooms would be harvested less frequently and that locals picked would collect fewer species in Hama-douri than in Aizu, hence the impacts of the accidents would be significantly lower in Hama-douri. However, these results showed that I certainly underestimated the impacts.

Unlike mushrooms, restriction on commercial use is applied separately to each species of wild vegetables. The Food Agency does not explain the reason for the difference. The government may have decided to more strictly control mushrooms due to generally higher radioactive Cs levels and higher variability among samples within a species population (Tables 4.9 and 4.10). In Aizu, it was

Table 4.10 Mean and standard deviation of radioactive caesium (134 and 137) in wild vegetable raw samples measured in Kawauchi Village in 2017 and 2018

Scientific name	*Radioactive Cs (Bq/kg)*
Aralia cordata	20 ± 38 (25)
Pteridium aquilinum	25 ± 21 (35)
Petasites japonicus	11 ± 15 (67)
Eleutherococcus sciadophylloides	854 ± 518 (6)
Aralia elata	212 ± 307 (13)
Parasenecio delphiniifolius	40 ± 50 (8)
Phyllostachys spp.[a]	68 ± 92 (30)

Sample size in parenthesis

Note:

a Three species of bamboo shoot.

only one of 11 popular species, on which 16 of 17 municipalities were subject to the government shipping restriction. That species is koshiabura *Eleutherococcus sciadophylloides*, whose new leaves are usually served as tempura (one of the nation's favourite dish). Kiyono and Akama (2013) found that the edible part, the young shoot, unfortunately, had higher levels of Cs than any other parts of the plant. The government prohibited commercial use of more species in Naka-douri and Hama-douri than in Aizu (Table 4.11). Not only Hama-douri but also the northern part of Naka-douri suffer high levels of radioactive contamination because the southeast wind changed to north, causing caesium particles to fall to the ground. Afterwards, the north wind contaminated the central and southern parts of Naka-douri, before the particles reached the north of Tokyo. The changes in wind direction generally caused the regional differences in the proportion of municipalities under the restrictions on commercial uses (Table 4.11).

While driving through the forested areas of Aizu, I observed no significant differences in the number of wild plant and mushroom pickers before and after the power plant accident. The survey from 2007 to 2010 counted 11 cars per 100 km while driving 444 km during the high seasons of wild plant and mushroom gathering. There were 12 cars per 100 km during the 372 km drive after the accident from 2015 to 2018. Some social changes occurred during the two periods. These are population decline as illustrated later, an increase in the number of outsiders and a change in regulation in which some communities ceased

Table 4.11 Percentage of municipalities in which commercial use of wild vegetables is prohibited

Scientific name	*Aizu*	*Naka-douri (north)*	*Naka-douri (middle)*	*Naka-douri (south)*	*Hama-douri (north)*	*Hama-douri (south)*
Matteuccia struthiopteris	0	100	50	7	63	60
Aralia cordata	0	0	10	0	63	60
Pteridium aquilinum	6	60	10	7	75	80
Petasites japonicus	0	20	10	0	38	20
Osmunda japonica	0	20	50	0	75	100
Hosta Montana	0	0	0	0	0	0
Eleutherococcus sciadophylloides	88	100	100	100	100	100
Aralia elata	6	80	73	40	88	80
Parasenecio delphiniifolius	0	0	0	0	0	0
Elatostema umbellatum	0	20	10	0	38	20
Sasa kurilensis	0	0	0	0	0	0

The nuclear power plant is located east of Nakadori (center) and near the border of Hamadori (north) and Hamadori (south).

exclusive use and discrimination against outsiders. These changes are interrelated. Before 2011, local people sold or gave as gifts roughly one-fifth of the harvests (Matsuura *et al.*, 2014). Such actions were no more possible after the accident, which must have significantly reduced incentives for harvesting. At the same time, local settlements suffered rapid population decrease, but they needed to patrol the forests to prevent outsiders from entering the land to exploit their resources. Some communities removed the signs of exclusive use, while others gave up patrols. As a result, after 2011, more pickers came from areas within one-day trip distance by car than before. Many of them came from cities south of Fukushima Prefecture, where population decline was not yet evident, and from areas east of Aizu, where contamination levels were relatively high.

Tadami broke the Japan's 24-h rainfall record in 2011, as mentioned before. It destroyed an 18 km forest road and changed the river course in the Kurotani Valley. Inland water fishery cooperatives stopped releasing juvenile fish into the Kurotani River and tributaries of the Tadami River till 2015. In addition to this damage, radioactive Cs levels exceeded the food safety standard for the two most popular species of fish, char and trout, in some areas of Aizu. Besides, rumours spread throughout the country, suggesting that it is dangerous to eat any kind of foods produced in Fukushima Prefecture. However, Cs levels declined (Figure 4.6). Anglers were pleased that the char had lower levels of Cs than the trout. They eat chars much more often, and their size was larger than trout. From 2015 to 2018, I totalled the number of cars to assess the abundance of anglers parked along two rivers in Tadami and one in Hinoemata. I counted 0.28 vehicles per kilometre, which was about a 10% decrease from the abundance measured from 2007 to 2010. I think such a decline would not be significant, considering the radiocaesium was still detected in the char and trout specimens from Tadami and Hinoemata between 2017 and 2019. The fishing sites were so attractive that most of them drove a long distance from the Tokyo metropolitan area. Caesium is supposed to stay in forest soils for many years, and

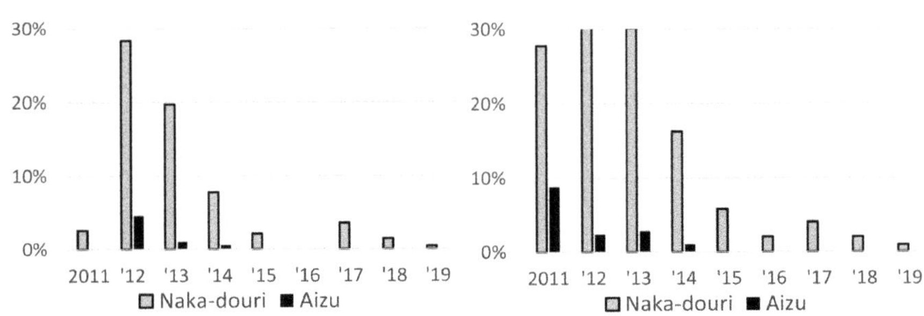

Figure 4.6 Percentage of small stream fish over the food safety standard (100Bq/kg of Cs-134 and 137 combined). White-spotted char *Salvelinus leucomaenis* (*left*), Yamame trout *Oncorhynchus masou* (*right*).

Source: Fukushima-ken[(4-10)]

river fish might uptake high levels of radioactive particles into the digestive tract. In contrast to Aizu, the levels of specimens brought to Kawauchi's testing centres in 2017 and 2018 were high, with an average radioactive Cs level of 92 Bq/kg for the trout specimens and 123 Bq/kg for char specimens. Over 40% of the samples exceeded the safety standard, and the inland water fishery cooperative has not resumed releasing juvenile fish.

Fukushima Prefecture reports the estimated annual number of visitors to various tourism sites, including hikers. Unfortunately, it does not select any location in the southwestern part of Aizu. Still, there are a few mountain peaks in the north-central part of the prefecture and some in the Abukuma Mountains relatively close to the power plant. Figure 4.7 shows that the 2011 incident had considerable impacts on the number of hikers in the two areas. This figure also contrasts the post-explosion trends. The decline in 2011 was below 25% of the previous year in Abukuma, a greater reduction than the 40% in the central mountains. Such a decrease was not a nationwide phenomenon since no similar decrease was found in the statistics for Nagano Prefecture, which has many mountains in the west of Tokyo (Shigematsu *et al.*, 2018). Although gradually recovering, the recovery rate was much lower in Abukuma, probably reflecting high levels of radioactivity (Sanada *et al.*, 2018).

On the other hand, I counted more cars parked at the foot of two mountains from 2015 to 2018 than before the accident, suggesting an increase in the number of hikers in Tadami. The number of tourists may have increased as well, as the number of visitors to the museum, Buna (beech) Museum, gradually recovered after 2011 to higher levels than 2010, the year it opened. The average annual number of visitors to the Tadami Dam Lake also increased after the accident. It was about a 50% increase, comparing before (2009 and 2010) and after (2016 to 2018). The lake has the most visitors in the town. Thus considerable differences are found in the recovery rate among different types of cultural

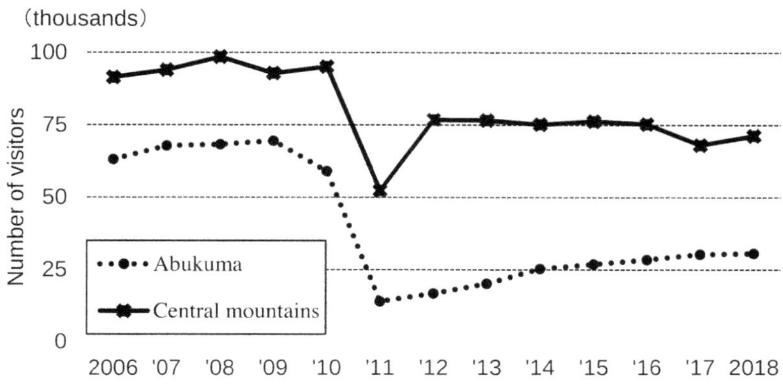

Figure 4.7 Change in the number of hikers in the Abukuma Mountains and the central part of Fukushima Prefecture.

services and regions. The impacts have been severe in the Abukuma Mountains and Hama-douri. Still, I do not see a significant decline in demands for such services as was the case for harvesting wild plants and mushrooms, angling and nature tourism in Tadami.

As the last topic in this section, I would like to mention bear hunting from two perspectives: damage control and use for food and medicine. Asian black bears *Ursus thibetanus* live in Honshu and Shikoku, whereas the other species, brown bears *Ursus arctos*, are found only in Hokkaido. Generally, the black bear populations are higher in the northern part of Honshu than the southern part and higher on the Sea of Japan side than on the Pacific side. Local populations in the westernmost part of Honshu are on the national Red List, whereas local people regard them as pests in the central and northern parts of Honshu. Most probably its population density is positively correlated with the amount of snowfall. Black bears occasionally encounter NTFP pickers and mountain hikers, injuring 93 people and even killing 1.2 a year on average from 2010 to 2019[(4-11)]. This risk is the major reason for control. Meanwhile, locals cook bear meat with vegetables as a Japanese type of stew, called kuma-nabe. Also, its gall bladder is very expensive as medicine, and hunters can sell it at ¥10,000 (ca. 92 US$) per gram dry weight.

The bear population density is high in Aizu and very low in Hama-douri, as 63% of the bears hunted for radioactivity inspection were from Aizu between 2011 and 2019. The five-year total of the number of bears killed in Fukushima, 1569 bears from 2012 to 2016, was the third-largest among the Japan's 47 prefectures. Considering the high proportion killed in Aizu, it is evident that the bear density in Aizu is one of the highest in the country. Radioactive caesium levels were high for the bears compared to most other foods harvested from forests, and 12% of meat specimens exceeded the safety limit in 2018 and 2% in 2019 even in Aizu[(4-10)]. The Ministry of the Environment reports (a) the number of black bears hunted for control and (b) the numbers killed during the hunting season, from 15 November to 15 February. The total numbers killed in the country in one year ranged roughly (a) 1000–4000 and (b) 300–500 during the years from 2007 to 2016. The statistics (b) reflect the hunter's own decision and willingness to acquire bears, which would represent levels of motivation for the hunters to obtain gall bladders and meat. The five-year total of the bears in (b) decreased only by 1% from 198 (2006–2010) to 196 (2012–2016)[(4-11)] in Fukushima Prefecture. I suspect that hunters are still eager for getting gall bladders despite the potential for high levels of radioactive Cs.

4.3.2 *Future population decline and forest use in Tadami*

Population decrease with rapid ageing is one of the most concerning issues in almost every rural area in Japan. Tadami Town marked the peak population, over 12,000 when the construction of the Tagokura Dam began in the 1950s. It was the largest dam in the country at that time and was completed in 1960. Afterwards, the population has continued to decline to nearly 4000 in 2020.

Further population decline will affect not only forest use but also community forest management. For example, local settlements and the local office of the national forest form a cooperative to determine how the national forests are to be used by people in the community within their administrative boundaries. Some communities do not allow outsiders to harvest NTFPs in the national forests in accordance with the agreement of this cooperative. There are signs at the entrance of national forest roads to prevent outsiders from using woods. Many people ignore the signs, so patrols are essential to avoid such conducts.

To estimate the future decrease in forest use, I obtained population statistics from the Tadami Town office, which covered ten years from 2009 to 2018, dividing the town into forty-one districts. I estimated the rate of birth, death, immigration and emigration in each district, based on sex-age-specific population. Next, I obtained each rate averaged across all the districts for each age group separated by five years (Table 4.12). There are a few points to be noted in terms of population sustainability. First, the birth rate is too low, and the population will decline even if emigrants do not exceed immigrants, as the total births per woman was estimated to be 1.81. Second, some young people emigrate when they become high school or university students, but many do not return. Third, the efforts of the town office to encourage immigration have not worked sufficiently. Although some programs invite young people from outside every year by offering supports for senior high school students with scholarships and funding for moving into a vacant house, the number of immigrants remain small.

I estimated population change based on a simulation model using STELLA (Varsity Wave). The results suggested that the town population as a whole would decrease to less than one-third of the 2018 level in 50 years. The percentage of people harvesting NTFPs varies by age group. Matsuura *et al.* (2013) divided the surveyed population into three age groups, i.e., 25–59, 60–79, 80 years and over, and found the proportion of those harvesting NTFPs was 25, 50 and 20%, respectively. The rate of decline was estimated to be a little bit greater among

Table 4.12 Average birth, death, immigration and emigration rates (per person per year) in 41 districts in Tadami Town

Age group	*Birth*	*Age group*	*Death*	*Age group*	*Migration (C)*	*Migration (P)*
20–24	0.059	65–69	0.009	15–19	−0.054	−0.056
25–29	0.130	70–74	0.010	20–24	−0.080	−0.128
30–34	0.108	75–79	0.023	25–29	0.042	−0.040
35–39	0.065	80–84	0.038	30–34	0.012	0.031
		85–89	0.082	35–39	−0.017	0.014
		90–94	0.157	40–44	−0.021	0.013
		95–99	0.281			
		100–104	0.604			

If the migration rate is negative, emigration exceeds immigration and vice versa. (C) shows the average of the central areas of Tadami, Asahi and Meiwa, and (P) shows the average of the surrounding areas.

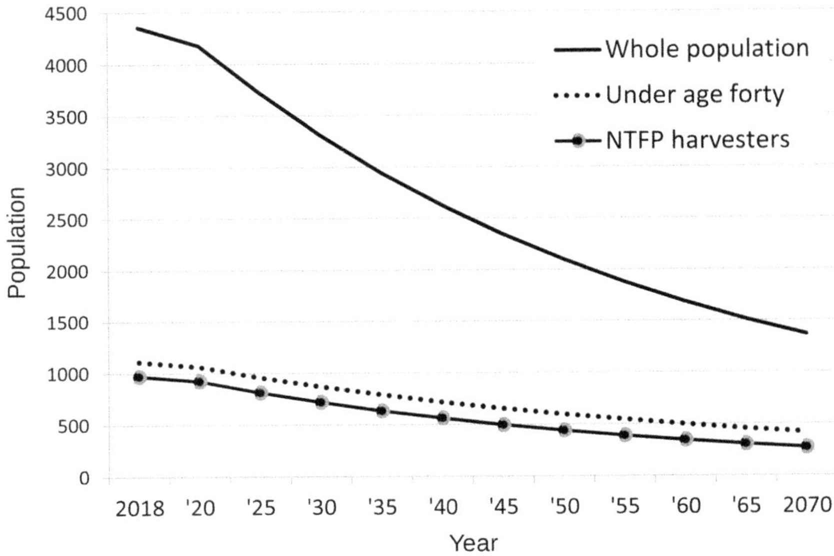

Figure 4.8 Estimated changes in the whole population, population under age forty and the number of NTFP harvesters in Tadami.

the NTFP takers than the whole population (Figure 4.8), because the effects of ageing are large. Small settlements are at risk of disappearing in the future. There were eight settlements whose population was no more than 50, and the estimation suggested that all of them would have no more than twenty residents by 2050, having no or at most a few people under age 40. Sooner or later these settlements will disappear, and their administrative district will be integrated into another, thus increasing the burden of forest management for already aged communities.

Tadami people said the nuclear power plant accident appeared to be the second blow, adding to the problems of ageing and population decline. The third factor might be more frequent and violent floods due to climate change. Every year heavy rains collapse small bridges across forest roads, blocking vehicle access (Photo 4.5). Neither the local government nor the Forestry Agency is financially able to repair such damages. On the other hand, global warming has reduced snow depth and considerably increased the number of boars, monkeys and deer. Of all the wildlife species, they are most damaging species to agriculture across the country (Chapter 6), while hunters have not only decreased in numbers but they are aged. Incentives for control hunting may have been reduced by contamination with radiocaesium and increased inconvenience caused by road damages. Such changes will probably not have significant effects on biodiversity conservation. The impacts of deer overpopulation are so great that they have limited ground vegetation in some parts of Japan. However, snow will still prevent such overpopulation for the time being. Demand for biomass fuel may

Photo 4.5 A concrete bridge across a small stream was broken into pieces in 2019, after another bridge had been destroyed in the previous year.

increase as mentioned in the previous chapter, but the reserve protects large areas of the national forests in Tadami. There are no obvious risk factors for biodiversity except climate changes. Decreased snowfall reduces soil moisture in summer, and possibly long periods of rainless days will negatively affect mushroom germination.

Supplementary notes

4-1) The Edo period lasted from 1603 to 1867. The impact of westernisation was minimal because of national policies that forced isolation from other countries. Western technology and culture were introduced into rural societies after the Meiji period (14868-1912).

4-2) Statistical yearbook published annually from 1956 to 2018 by Rin'ya Kōsai-kai.

4-3) The Forestry Agency had built many roads in the national forests, relying on the debt from the Ministry of Finance before the agency abandoned the independent profit system. The town government is dependent on central government's tax revenues, and it cannot afford to make them available for building forest roads because of a decreasing amount of the revenues due to the agency's financial deficit.

4-4) Family Cercidiphyllaceae has only one genus and two species in the world, living in China, Korea and Japan. One of them, *Cercidiphyllum japonicum*, is found widely from Hokkaido to Kyushu in Japan (Department of Biosphere-Geosphere Science Faculty of Biosphere-Geosphere Science Okayama University of Science). Available at: http://had0.big.ous.ac.jp/index.html (accessed 12.12.19).

4-5) *Annual Checklist Interface v1.9 rev 2126ab0*. Available at: https://www.gbif.org/species/2857132 (accessed 20.12.19).

4-6) *English names for fungi (July 2019)*. Available at https://www.britmycolsoc.org.uk/library/english-names (accessed 20.12.19).

4-7) Naikaku-hu Tōkei-kyoku conducts a national population census every ten years, which is regarded as the most reliable statistical source.

4-8) These days vehicle registration plates cover smaller areas than before, occasionally confined to only one city whose population is about 20,000.

4-9) Fukushima Hukko Centar. Available at: https://www.pref.fukushima.lg.jp/site/portal/ (accessed 9.12.20).

4-10) *Fukushima-ken: nōrin-suisan-butsu kakō-hin monitoring jōhō*. Available at: https://www.new-fukushima.jp/result (accessed 3.01.20).

4-11) *Kankyō-shō, hokakusū oyobi higai no jōkyō*. Available at: https://www.env.go.jp/nature/choju/docs/docs4/index.html (accessed 16.01.20) Local people submit hunting permit applications for wildlife harmful to agriculture and human life, and hunting is carried out after specific types of damage is identified or for population control purposes to prevent excessive harm. The ministry of the Environment publishes the statistics of these two types of management separately.

5 Forests and satoyama landscapes in the suburb of a metropolitan area

5.1 Disappearance of traditional satoyama and current conservation activities

Many articles on satoyama landscapes deal with biodiversity and ecosystem services (Takeuchi *et al.*, 2003; Duraiappah *et al.*, 2012). Kuramoto and Sonoda (2003) and Akimichi and Morimoto (2012) indicated that there were more species to be protected in secondary forests than climax forests. Some articles (Takeuchi, 2003; Saito and Shibata, 2012) seem to argue that coppice management is essential for conserving high levels of biodiversity. I call the authors of these articles satoyama advocates in this chapter. Sometimes their argument does not take an appropriate data-driven approach to the relation between biodiversity and ecosystem services. In other cases, they fail to have a broad enough geographic view and their argument applies only to a limited area and is not applicable to other areas in Japan. My perception is different. First, it is not reasonable to compare secondary forests with more intact ones, insisting that the former is more important than the latter. Second, coppice management is not common practices and, in my view, is not essential for conservation. A large patch of beech forest has many species that are not found in satoyama, as I indicated in the last chapter. Likewise, satoyama does not have many endemic and endangered species like Amami Oshima. To examine the validity of such arguments, I monitored leisure activities and bird populations for several years to find any relation between leisure activities and levels of biodiversity. I selected landscapes in Tsukuba City as an example which is about 50 km north of Tokyo. The study area was a typical satochi landscape[5-1]. As explained later, my conclusions are as follows: first, high species richness is neither a product of satoyama conservation activities nor of landscape design policy that takes biodiversity into account; second, the most prominent contribution would be the opportunities to have a pleasant time for the residents living nearby.

It would be true in the south-western part of Japan that satoyama has the richest biodiversity, where okuyama (remote mountains) is predominantly covered by young even-aged tree plantations of single species, as described in the following chapter. Satoyama has many more naturally regenerated secondary forests than okuyama, associated with rice paddies and occasionally with an artificial pond used to supply water for irrigation. Such a variety of ecosystems can hold more species than plantation forests. The majority of ecologists work in large cities in Honshu and Kyushu, where satoyama landscapes are not very far from their offices or laboratories. In contrast, such regions as Amami and Tadami are too remote for them to visit as often as to produce high-quality scientific papers. Therefore, not many universities have a group of ecologists who study ecology and conservation of okuyama or remote islands. As a result, they probably cannot help arguing that satoyama has representative and richest Japanese biodiversity. I have read some articles in which ecologists asserted that "biodiversity would be lost" if satoyama forests were left unmanaged. The Japanese government agreed with such arguments and coined a term, satoyama Initiative, as a catchphrase, when it held the Biodiversity COP in Aichi in 2010.

I recognise two conceptual ideas of satoyama among scientists. One perceives traditional coppice woodland as an essential element of the landscape for biodiversity conservation (Takeuchi, 2003; Saito and Shibata, 2012). Yet remaining coppice forests are too small and too fragmented to maintain a representative set of species in the relevant region. There are indeed many satoyama conservation activity groups in the country. Still, most of them are not interested in restoring coppice forests, as exemplified later by a list of such groups and their objectives in Kanagawa Prefecture. Even if they are involved in coppice wood management, I suspect that forested areas are too large for them to have their efforts make a significant contribution to biodiversity conservation, as the papers cited above maintain. In this section, I provide some information on satoyama activities and criticism based on Kawano (unpublished) and additional databases. Reflecting these analyses and discussions, I suggest that other regions should play more critical roles in Japanese biodiversity conservation: remote islands of the Nansei and Ogasawara Archipelago as well as okuyama. I give some lists of vertebrate species endemic to Japan and their endangered status in the last chapters. The other idea is that it is a landscape mosaic that maintains high levels of biodiversity, which holds various types of ecosystems such as forests, grasslands, ponds, ditches and paddy fields within a limited area. It is a product of human activities that have created a small-scale landscape that maintains high species diversity in Japanese rural regions (Katoh *et al.*, 2009). I agree that traditional management practices would be better for biodiversity than the current practices, but it is not practical to employ them in large areas. My argument in this chapter employs the second idea, a combination of various landscape elements rather than traditional management.

A traditional satoyama typically had coppiced forests of oak *Quercus serrata* and *Quercus acutissima*. *Q. serrata* does not grow well in the lowlands of Kyushu, where evergreen trees predominate. Local people used fallen leaves of the oaks to produce fertiliser. Moriyama (1997) estimated that such coppiced forests covered about one-third to a quarter of satoyama in Tsukuba until sometime in the 1960s. Trees were cut for charcoal and firewood typically within 30 years after the previous cut. A much larger area of grasslands was maintained than these days because people had horses, cows and goats to feed and covered their roofs with grasses, such as *Miscanthus sinensis*, *Imperata cylindrical* and *Phragmites australis*. Rice paddies kept water nearly all the year round as well as artificial ponds supplying water to the paddies. All these landscape elements provided habitats suitable to spring ephemerals, grassland species and aquatic plants and animals.

There were drastic and fundamental changes in satoyama landscapes after charcoal and firewood had been replaced by coal and petroleum, and leaf manure by chemical fertiliser in the 1960s. Most of the grasslands disappeared because there was no need to feed animals and no roofs to cover with grasses. Rice paddies are not flooded anymore, except for the growing season from May to sometime in summer, and are left dry until the following planting season. Many artificial ponds disappeared after fine networks of artificial waterways were developed. At the same time, provisioning services nearly disappeared from satoyama forests and grasslands. Coppice forests were abandoned, and they were succeeded by forests with dense undergrowth and larger trees, or replaced by even-aged single-species tree plantations. Farmers prefer to use herbicide on small paths in their farmland, causing further decline in the herbaceous species diversity.

Such great changes in rural landscapes must have promoted a variety of activities in which many groups are involved. The Ministry of the Environment gives a list of such groups on a website. Mizuki Kawano, one of my students at Nagasaki University, looked up the website of each group, and examined the activities that those groups were involved in and classified the activities as follows: (a) enjoying nature observation; (b) thinning; (c) clearing underbrush and weeding primarily for scenic beauty; (d) clearing bamboo groves to be replaced by natural forests; (e) producing charcoal and firewood; (f) clearing fallen leaves not for producing manure but probably as a scenic choice; (g) mushroom production on logged oak trees; (h) various experiences in environmental education programs for children; (i) planting deciduous trees that they prefer for scenic beauty after cutting an unmanaged and abandoned forest; and (j) biological survey.

Many private forest owners are not concerned with forest management these days since they do not think forests deserve management for timber production. The activities (a) and (j) originated from interests in biodiversity. The two types of silvicultural activities, (b) and (c), contribute to developing suitable habitat for spring ephemerals, and activity (d) will result in higher levels of species diversity. The activities (e) to (g) are relevant to traditional products in satoyama. As such, most of these activities are related to conventional satoyama forest management. What is lacking is landscape perspective, and these groups' interests are mostly limited to forest ecosystems without paying attention to other components of satoyama.

There were 972 satoyama groups in a list provided by the Ministry of the Environment in 2001[(5-2)]. There were many groups, about one third, in the most populated seven prefectures in the Tokyo and Osaka metropolitan areas. The number of participants per group was greater in the 7 prefectures than the other 40, and it seemed that the total number of people who joined the activities there was about the same as that of the other 40 prefectures. Forested area per population is small in these seven prefectures, especially in Tokyo and its vicinities, suggesting that a small patch of forest is precious for many people in urban areas. It is also true of Tsukuba City, about 50 km away from Tokyo, and small patches of forest are targeted by satoyama activists as described later. In contrast, there are no satoyama conservation groups in Amami and Tadami, while there are some nature conservation groups that are ready to fight against forest cutting.

Kanagawa Prefecture gives an excellent example that helps understand the relation of recent satoyama activities with the restoration of traditional landscapes. The prefecture is west of Tokyo and the number of groups was as large as Tokyo (Kawano, unpublished). It has the second largest population among all the prefectures and 947 km^2 of forests, covering 39% of the land. Most of the forests are located in the west and north, while forests also cover hilly areas of a small peninsula in the southeast. There were 58 groups in 2001. A total of 24 worked in the woods on the peninsula and adjacent forested areas, and 17 worked in the north and west. Most of them participated in forest management and more than two activities among (a) to (j) as classified above[5-3]. Table 5.1 gives the number of groups involved in each activity. Environmental education was the most popular, followed by nature observation, and 12 groups practised both of them. A total of 24 groups practised thinning, the primary purpose of which was to encourage the growth of residual trees, but it also must have improved scenic beauty within forests. It is a silvicultural treatment to increase tree size rather than coppice management. The results suggest that the primary objectives are irrelevant to the restoration of traditional satoyama forests in most of those activities. In addition, they seemed to have worked in a

Table 5.1 Number of organisations involved in various satoyama activities in Kanagawa Prefecture

Kind of activity	*Number of groups*
(a) Enjoying nature observation	26
(b) Thinning	24
(c) Clearing underbrush and weeding primarily for scenic beauty	16
(d) Clearing bamboo groves to be replaced by natural forests	12
(e) Producing charcoal and firewood	13
(f) Clearing fallen leaves probably not for producing manure but scenic preference	9
(g) Mushroom production on logged oak trees	2
(h) Various experiences in an environmental education program for children	31
(i) Planting deciduous trees for scenic beauty after cutting an unmanaged and abandoned forest	12
(j) Biological survey	9

Source: Kawano (unpublished).

small area, less than a hectare of the forest as far as I observed such activities in Tsukuba. The number of groups was too small to take care of the forests in the northern and western parts of the prefecture, where biodiversity levels are higher than the peninsular in the southeast. Seventeen groups had a combination of (a) nature observation program and silvicultural or production activities, (b) to (e). These facts imply that the most important objectives were education for children through biodiversity conservation. These activities provide little contribution to restoring traditional landscapes. In the following sections, I emphasise that satochi and satoyama landscapes would be able to maintain rich species diversity even if forests are left unmanaged. I also argue that the provision of leisure spaces would be an important ecosystem service than biodiversity conservation, in these landscapes, in contrast to Amami and Tadami.

5.2 Focusing avian species diversity in Tsukuba

The area of Tsukuba City is 284 km^2 and has Tsukuba Mountains on the northern border, whose highest peak is about 700 m above sea level. The rest of the area has fragmented woodlands surrounded by farmlands and residential settlements on a flat terrain, recognised as exceptional landscapes in predominantly mountainous Japan. The national government built Tsukuba Science City in 1970, in an area where agricultural landscapes dominated, and moved most of its national research institutes into the city. The landscape, so-called satochi, is typically composed of rice paddies, dry fields, woodland, small streams and residential areas (Photo 5.1). Ecosystem composition is fundamentally the same as satoyama, while the proportion of farmland to the forested area is larger in satochi than satoyama. The city's population increased nearly fourfold from 1973 to 2015, after the research complexes were built in the city. The city design separated residential settlements from each other so that they were next to satochi

Photo 5.1 Satochi landscape in Tsukuba City.

landscapes, and the residents would be able to enjoy a rural atmosphere. Along with the development of the Science City and population increase, there were considerable landscape changes. Wooded areas declined and became more fragmented from 18.8% of land cover in 1972 to 14.4% in 2012[(5-4)]. Likewise, the dry field land area decreased from 34.9% in 1976 to 30.9% in 1991 (Sugimura *et al.*, 2007) and has declined further since then. There has been only a slight decrease in rice paddy areas over the past three decades.

I set up a study area in Tsukuba, where I recorded bird and herbaceous species abundance along survey routes. Also, I counted the number of people visiting for leisure, as described in the next section. Census routes had a total distance of 48 km, and they were in the southern part of the city along three small rivers, the Ono, Inari and Higashiyata. Next to the rivers, there were rice paddies, dry fields, woods and residential districts, generally located away from the river in this order. There were patches of broadleaf trees on narrow strips of slopes, situated between farmland and residential areas. Demand for building houses on the slopes was not high, and the landowners were not very willing to plant timber trees so that secondary forests remained on most of the slopes. The river basins had a small range of elevation, with bases that were 10–20 m and flat plains above the slopes around 25 m above the sea. No traditional landscape management such as coppicing, leaf manure production and maintenance of hay fields, occurred. Coppiced forests, composed of *Q. serrata* and *Q. acutissima*, used to cover much more extensively and were utilised as a source of fuel and manure (Sprague *et al.*, 2000). Most of these forests had been left unmanaged, and the deciduous oaks were replaced with evergreen trees, such as *Quercus myrsinaefolia*, often associated with a dense undergrowth of bamboo grass *Sasaella ramose*. There was an artificial pond for irrigation and a small patch of grassland for cattle breeding research along the survey routes.

I divided the whole area into three: (A) North Ono Basin, (B) South Ono Basin and (C) Inari and Higashiyata Basin. Table 5.2 gives the proportion of land cover types along the routes. Tree plantations of *Cryptomeria japonica* and *Chamaecyparis obtusa* also have expanded as well as evergreen broadleaf forests since coppiced forests were abandoned. These forests and tree plantations had variable stand structures. Volunteer groups occasionally thinned and weeded some of them, subsidised by the city government, while some unthinned *Chamaecyparis* plantations had little undergrowth vegetation. Small streams had either tall grass on the bank or concrete walls with low vegetation. Area (A) had

Table 5.2 Percentage of land use types along the census routes

Area	*Secondary forest*	*Tree plantation*	*Chestnut or plum orchard*	*Bush*	*Rice paddy*	*Dry field*	*Stream or pond*	*City park with trees*	*Residential area*
(A)	1.1	4.4	1.6	7.2	26.8	22.1	0.1	6.2	30.5
(B)	5.5	5.1	2.9	8.8	32.1	21.6	3.5	0.1	20.3
(C)	8.3	1.5	0.5	5.6	30.2	8.1	0.2	0.0	4.1

Photo 5.2 Urban park near a residential complex in Area (A).

housing complexes associated with patches of trees in city parks, which was not representative of a satochi landscape (Photo 5.2). The significant difference from satoyama is that woodland covers a relatively small proportion of the landscape. An exception was Takasaki Woodland in Area (C), where the city government maintains it as a forest park to attract visitors, cutting undergrowth regularly for scenic beauty (Photo 5.3).

Photo 5.3 Takasaki Woodland: thinning and clearing undergrowth are practiced, while it does not resemble a traditional satoyama forest due to absence of coppice management.

There is nothing very particular about mammals in the satochi study area, in comparison with Amami that has several endangered species endemic to the Nansei Archipelago (Table 2.1) and Tadami that has many species endemic to Japan as well as large mammals (Table 4.3). Hare *Lepus brachyurus* and an introduced species of civet *Paguma larvata* were the largest mammals I have seen in satochi landscapes in Tsukuba. In contrast, I found the landscapes were rich in avian species, reflecting a wide range of habitat types. I employed route census to save time and count human visitors at the same time, instead of point-count census employed in the other study areas. Finding birds from a bicycle was easy in rice paddies and dry fields. Table 5.3 lists the bird species that I heard or sighted in the morning after sunrise during the breeding season. I observed them repeatedly within a small range of locations so that it was sure that they were not transient.

The result suggests that the satochi in Tsukuba had a few differences from Amami and Tadami. First, the number of species observed, i.e., 36, was large relative to the size of the census area despite highly disturbed habitats. The study area extended north and south for about 9 km and east and west for about 4 km. Tadami had the 97 breeding species in large areas of okuyama and satoyama, which extends north and south for about 20 km and east and west for about 27 km (Table 4.4). I counted 21 breeding bird species in Amami, 1985–2017. The study area was about equal to that of Tsukuba, extendeding 12 km. Second, only two species of them were regarded as endangered, grey-faced buzzard *Butastur indicus* (VU) and plumed egret *Egretta intermedia* (NT). Katoh *et al.* (2009) described the buzzard as a symbolic endangered species of satoyama and representation of high species diversity. "High species diversity" is an appropriate expression, but the meaning of "symbolic endangered species" is not very clear. Symbolic species always appeared whenever conservation groups attempted to protect natural ecosystems from development activities in Japan. The national government constructed a highway passing through rice paddy areas in Tsukuba and cities nearby. There occurred an environmental protest, but the symbolic species was not the buzzards but bean goose *Anser fabalis*. The goose had winter habitat more than 2 km away from the highway, while the raptor used the paddies along the highway as parts of feeding sites. Third, one-third of the species are dependent on the wetland, reflecting the fact that rice paddy was a crucial component for sustaining diversity. The birds stay in the paddies when flooded. They fly to the south after September harvest and come back when paddies are flooded in May. Species diversity is relatively low when the paddies are dried, as I see only one migratory species feeding in the paddies in winter, *Turdus eunomus*. Fourth, all the species, even the two species in the Red List, have a large distribution area, and Tadami also recorded all of them but for a species of egret, *Egretta intermedia*. Many buzzards winter in Amami, where conservationists do not pay much attention to them.

Now that forest area is small in satochi, the birds in Table 5.3 may fail to include some satoyama species living in forests. So I surveyed forest birds on north-facing slopes of Mt. Tsukuba to examine if the area had any species not found in

Table 5.3 Breeding birds observed in satochi landscapes in Tsukuba

Scientific name	*Common name*	*Major habitat*[a]	*Migration pattern*[b]
Butastur indicus	Grey-faced buzzard	F, W	M (J)
Accipiter virgatus	Japanese sparrow hawk	P	M (T)
Bambusicola thoracica	Chinese bamboo pheasant	F	N
Phasianus colchicus	Green pheasant	G	N
Streptopelia orientalis	Eastern turtle dove	F	N
Cuculus canorus	Common cuckoo	P	M (T)
Cuculus poliocephalus	Little cuckoo	F	M (T)
Alcedo atthis	Common kingfisher	W	N
Dendrocopos kizuki	Pigmy woodpecker	F	N
Hirundo rustica	Eastern house swallow	W	M (T)
Motacilla alba	Japanese pied wagtail	G	N
Motacilla grandis	Japanese wagtail	W	N
Hypsipetes amaurotis	Brown-eared bulbul	F	N
Lanius bucephalus	Bull-headed shrike	F	N
Cettia diphone	Japanese bush warbler	F	N
Acrocephalus arundinaceus	Great reed warbler	G	M (T)
Parus major	Japanese great tit	F	N
Aegithalos caudatus	Japanese long-tailed tit	F	N
Zosterops palpebrosa	Japanese white eye	F	N
Emberiza cioides	Japanese meadow bunting	F, G	N
Chloris sinsia	Small Japanese greenfinch	F, G	N
Passer montanus	Japanese tree sparrow	G	N
Alauda arvensis	Japanese skylark	G	N
Sturnus cineraceus	Grey starling	G	N
Corvus levaillantii	Japanese jungle crow	F	N
Corvus corone	Eastern carrion crow	G	N
Cyanopica cyanus	Japanese blue magpie	P	N
Garrulus glandarius	Japanese jay	F	N
Arden cinerea	Eastern grey heron	W	N
Nycticorax nycticorax	Night heron	W	N–M (J)?
Ardea alba	Great egret	W	M (J)–(T)?
Egretta intermedia	Plumed egret	W	M (J)–(T)?
Egretta garzetta	Little egret	W	N
Bubulcus ibis	Indian cattle egret	W	M (J)
Charadrius placidus	Long-billed plover	W	M (J)
Anas Poecilorhyncha	Spot-billed duck	W	N

Notes:

a F: forest; W: wetland, rice paddy; G: grassland, dry field; P: urban park with a variety of broadleaf trees planted for scenic beauty.

b M (J): migrate to the southern part of Japan and further south; M(T): migrate to the tropics; N: non-migratory.

fragmented forests of satochi. I studied from 1999 to 2003 and from 2006 to 2007. The mountain range has the highest peak in the city, 877 m above the sea. Forests extend 23 km from the southwestern base of Mt. Tsukuba to northeast with 4–7 km width of mountain slopes. The other mountain peaks range from 400 to 700 m high, and secondary forests of broadleaf trees and conifer plantations cover their slopes. I set up eight census points and counted birds on the northern slopes of Mt. Tsukuba from the elevation of 190 m near the forest edge to 815 m. An old map shows that three tree species partially covered the slopes, pine *Pinus densiflora*, cedar *C. japonica* and cypress *C. obtusa*[5-5] in the late 19th century. Pines were at the same position with the current forests that have a larger area than before up to about 300 m high. Pines replaced the cypress trees and the current plantations are at around 400–500 m in elevation. There were also oak (*Q. serrata*) forests that must have been coppiced and utilised by local people living down in the valleys below. Pine forests were left unmanaged and later mixed with naturally regenerated oaks and other broadleaf trees. Broadleaf forests were young at the time of bird survey, which had been cut clear once for pulpwood production (N. Kawaji, personal comm). Composition of dominant tree species changed gradually toward the mountain top, where forests above 750 m had been protected from cutting[5-6] and dominated by *Fagus crenata*, *Quercus crispula* and hornbeams *Carpinus* spp. I avoided counting birds in the plantations of *C. japonica* and *C. obtusa*, because species diversity was low there.

I never used the term satoyama in the last chapters on Amami and Fukushima, because satoyama landscapes do not play an indispensable role for biodiversity conservation. In this section, I attempt to find differences in avian species composition between satoyama and okuyama type of forests. I found 32 species of birds in total, excluding a recently introduced species, *Leiothrix lutea*. It was introduced from China, sold as a pet, and unsold birds were released in the mountain area. The introduced population had not been controlled, although it is on the List of Alien Species in Japan according to the relevant law. They were ranked third in the number of individuals counted per monitoring plot, following bulbul *Hypsipetes amaurotis* and warbler *Cettia diphone*. The bulbul was also one of the most frequently counted birds in Amami, in young secondary forests in particular (Figure 2.8). The woods around the three lower elevation plots were much closer to farmlands, while the forests around the four higher plots were much less disturbed than the others. Excluding the fourth point near tree plantations, I divided the woodlands into two groups; one around the lower three points and the other around the upper four. There were 16 forest-dependent species not found in satochi, of which six species were not found at lower altitudes. Probably they selected habitats with cooler climate conditions. I thought of two factors as the reason for the absence of the rest of ten species from satochi; lack of many large trees and small fragmented forests. Four of the ten species are cavity nesters and one of them I found more abundant in old forests than young ones in Amami. These five species are likely dependent on large trees. The rest of the five species may be dependent on relatively large patches of forest.

According to the division of the forests, I was able to divide the birds into two groups; (A) found more frequently at lower altitudes than at higher altitudes and (B) those that appeared to prefer living at higher altitudes, supposing that (A) represented satoyama type of birds and (B) okuyama type. Group (B) had six more species than (A). Group (A) had thirteen species of which three were not found at higher altitudes (crow *Corvus corone*, magpie *Cyanopica cyanus* and pheasant *Bambusicola thoracica*). These species may be typical satochi-satoyama species. The pheasant was introduced from China and released repeatedly since the 1930s for hunting[5-7]. Two of the 13 species were not found in the satochi survey (grosbeak *Eophona personata* and flycatcher *Ficedula narcissina*). The flycatcher was found more frequently in old forests than young forests in Amami, and it was one of the most common forest birds in Tadami. Recently, during the breeding season in 2019, I found the flycatcher in three patches of satochi forest in Tsukuba, when trees had grown bigger after the previous survey period. Thus the grosbeak is the only species that I thought required relatively large patches of satoyama forest. The number of species in satochi forests was not significantly different from that in the group (A) above. Therefore, there is no reason to believe that satoyama forests have richer species diversity than okuyama, while okuyama may have a wider variety. As the study in Amami suggests, many bird species prefer to live in older forests with large trees than younger ones with small trees (Table 2.4). It was most probably habitat variation from the foot to the top that maintained the species diversity in the case of Mt. Tsukuba. satoyama advocates argue that traditional and active forest management is necessary to have species richness. However, as far as avian diversity is concerned, it should be noted that a variety of ecosystems located next to each other contribute to maintaining species richness.

The prefecture museum, Ibaraki-ken Shizen-hakubutsukan (1998), surveyed birds and mammals, giving lists of species found in field observations on the southern and northern slopes of Mt. Tsukuba, based on scientific reports and interview surveys to local people on sightings and field signs. In addition to the breeding bird species I identified, they found three species below the elevation of 300 m; wagtail *Motacilla cinerea*, minivet *Pericrocotus divaricatus* and buzzard *Pernis ptilorhynchus*, in 1995 and 1996. They found four species in forests at a higher elevation, in addition to the species identified in my study. Then I compared the list of birds found in Tsukuba and that in Tadami (Tadami-chōshi Hensan-iinkai, 2001), and found eight species breeding in Tsukuba but not in Tadami except for introduced species; such as *Egretta intermedia*, *Falco columbarius*, *Phasianus versicolor*, *Gallinula chloropus*, *Scolopax rusticola*, *Ninox scutulata*, *Hirundo daurica* and *Erithacus akahige*. Only one of them, *Erithacus*, was found at higher altitudes and the others were found either in farmlands or forests in satochi. In contrast, 15 species of breeding birds were found in forested areas of Tadami and not in Tsukuba. These findings in Tadami and higher elevation of Mt. Tsukuba suggest that okuyama (remote mountains) has many avian species not found in satoyama landscapes.

Mammal surveys are more difficult and time-consuming because they are often nocturnal, and researchers use a variety of techniques depending on size and behavioural characteristics. The museum identified 23 species in the same Mt. Tsukuba area with such efforts; shrew, two moles, three bats, hare, three squirrels, four mouse, rat, vole, racoon dog, fox, three weasels, civet and boar (Ibid). Fifteen of them were found through using field signs and/or trapping. The number of mammal species endemic to Japan, i.e., 11, was half the number found in Tadami, including a species of vole, *Microtus montebelli*, which was not identified in Tadami. Another difference from Tadami is that no large mammals live in Tsukuba, such as bear, deer and serow, now that the forests are isolated from the distribution areas of these species. As the last point to be noted among the vertebrate taxonomic groups, a unique species of salamander live in a minimal range on the slopes of Mt. Tsukuba. Taxonomists separated this species from the other species of genus *Onychodactylus* and named it as *O. tsukubaensis* in 2013. Then Ministry of the Environment determined to protect it under the Act on Conservation of Endangered Species of Wild Fauna and Flora since 2015. However, its habitat area remains unprotected outside the quasi-national park zones mentioned above. It is one of a few species living in Honshu, which are ranked high in the national Red List (Table 1.2). Hayase and Yamane (1982) found its juveniles in small streams above 500 m high, indicating that its survival does not depend on satoyama landscapes.

5.3 Leisure use and conservation value of the landscape

Many studies emphasised the significance of scenic beauty that Japanese rural landscapes provided in local people's daily lives. Some of them indicated that such landscape components as rice paddies, dry fields and forests were given a higher evaluation by residents than the other parts when they showed rural landscape photographs (e.g., Matsumoto *et al.*, 1999; Tanokura *et al.*, 1999; Okada and Asakawa, 2001). Other studies evaluated recreation sites concerning travelling time and population size of the cities that the visitors lived in (Aoyagi, 1991; Kunimitsu, 2005). These studies were based on questionnaire surveys and suggested that the frequency of their visit to satochi had a negative correlation with the distance they travelled.

After I moved to Tsukuba in 1997, I noticed that many people visited satochi for leisure. I was curious if frequently visited sites had any particular landscape features, and started field observation to examine the visitors' behaviour. Also, I attempted to obtain practical implications on landscape management from this survey. I bicycled and sighted visitors on some routes between home and the work office, during the period from May 1999 to March 2006, and recorded locations, activity patterns, their sex and age class roughly estimated from outward appearance. The census was conducted in the morning and before dark on the way back home. I also moved around during the weekend. Based on the frequency of each activity observed and laboratory works, I developed a database in which I summed up the number of visitors after dividing activities into several types, weekday and weekend as well as morning and afternoon. I also incorporated location, land use types and population size of each residential district into the database. Then I developed a use frequency index, calculated its values and used the index as the dependant variable in linear regression models to find what kind of landscape elements (independent variables) had a significant relation with leisure use.

Dividing the study area into three, (A), (B) and (C) (Figure 5.1), the aggregated population size was about 15 thousand, 7 thousand and about 13 thousand

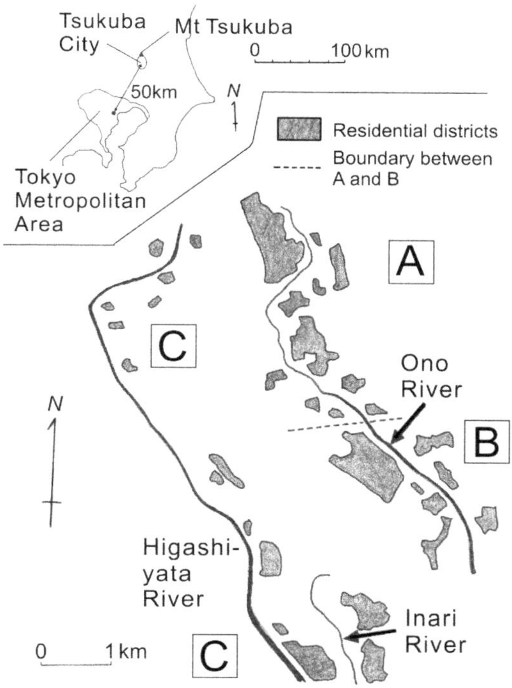

Figure 5.1 Study area and location of Tsukuba City.

within a distance of 2 km from the census routes, respectively. There were 50 residential settlements with an average population size of 699. Area (A) had the largest population and the highest density of the three. (B) was mostly under agricultural use with scattered residential settlements except for one with a population of about 1600. (C) had relatively large woodlands and residential districts in the southern part. I divided the census routes further into 150 sections and took the average number of visits per kilometre per census for each section to compare the use frequency among the sections. Then I examined if there would be a significant relationship between the average (use frequency) and landscape factors, using simple linear regression models. I considered such factors as land use types, population size, the distance between the visited site and residential settlements and the number of bird species.

I counted a total of 3900 people having leisure time along the census routes. In terms of activity patterns walking and dog walking accounted for more than 90% in all three areas (Table 5.4). The latter was more frequently observed, particularly in Area (B), where agricultural land was the most prevalent (Table 5.2). Walking was most commonly observed in Area (C), where secondary forests and residential districts were close. Recreation was observed mostly among children collecting insects or aquatic animals, riding a monocycle, roller skating, etc.

Table 5.4 Patterns of use observed in the study area: percentages and use frequencies (in parentheses)

Area	*Frequency of survey*	*Walking*	*Dog walking*	*Jogging*	*Recreation*	*Total*
(A)	375	34% (405)	61% (720)	2% (27)	2% (27)	(1179)
(B)	372	30% (401)	64% (856)	3% (40)	3% (40)	(1337)
(C)	112	52% (718)	43% (593)	5% (63)	1% (10)	(1384)

The purpose of walking includes exercise and relaxation.

I often watched people fishing along a river, but counted only a small number of them because their locations were not on the survey routes. Crude estimates by age group suggested that those in their sixties were observed most frequently. The group had retired from full-time jobs within 10 years in general at that time, although not nowadays due to rapidly ageing society. The number of visitors on a weekend afternoon was about 50% larger than weekday afternoon and three to four times than weekday morning.

The ten most frequently visited sections among 150 had a certain pattern in the landscape. They included high percentages of rice paddies and secondary forests on the average, 49 and 21%, respectively, relative to the means of the whole routes (Table 5.2). Particularly notable was that the percentage of secondary forests was much higher than the average. Besides, seven of them had a relatively large population, over one thousand, within a distance of 1 km (Photo 5.4 as an example). I sighted 4.3 persons per kilometre per census on weekends as the highest average value of the number of visitors, at least one person every time I visited. In contrast, people tended to avoid passing through residential areas, so the ten least frequently visited sections contained a high proportion of residential areas along the roads, 44% on average. Regression analyses suggested that the most critical landscape

Photo 5.4 Most frequently visited section through paddy fields and secondary forests in the vicinity of a residential district.

element (most predictive independent variable) was a combination of rice paddy and secondary forest (Sugimura, 2007). In other words, people preferred visiting the routes that ran through between a rice paddy and a secondary forest.

One of the linear regression models that supports the statement above is

$$Fi = 0.39R{*}Wi - 0.16Lif - 0.21Lir + 0.35\Sigma Pik/(Dik\hat{}1.3)$$

Fi: use frequency index for section *i*

*R*Wi*: percentage of section *i* that ran between rice paddy and secondary forest

Lif, Lir: percentage of section *i* that ran along dry field and residential district, respectively

Σ*Pik/*(*Dik*^1.3) denotes a population gravity parameter, in which ***Pik*** is population size of settlement ***k***, ***Dik*** is distance from the centre of ***k*** to section *i* and Σindicates that *Pik/*(*Dik*^1.3) is summed up along section *i* as described in the text below.

The power 1.3 associated with ***Dik*** followed Aoyagi (1991), and *Pik/*(*Dik*^1.3) performed better than *Pik/Dik* as a population gravity parameter in regression analyses. The coefficient values are standardised to measure the effect of an independent variable on the dependent variable. The coefficient values, when compared with each other, suggest that a combination of rice paddy and secondary forest might have a stronger effect than population size along the route nearby.

Bird species diversity also had a significant positive correlation with variable *Fi* as well as with *R*Wi*. The number of bird species counted had a positive correlation with the area of secondary forest (Figure 5.2). Takasaki Woodland and a piece of forest nearby across a paddy field had a few species that were not

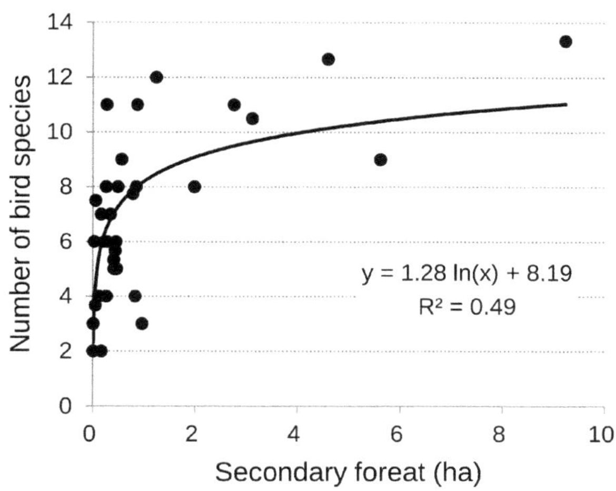

Figure 5.2 Relation between bird species richness and area of secondary forest along census route.

common in the other woods. They were *Garrulus glandarius*, *Aegithalos caudatus* and *Zosterops palpebrosa*, which I located at more than half the census points in the Mt. Tsukuba area. This area attracted many visitors, although it was not close, about 1 km, to a settlement with a relatively large population. These findings suggest that secondary forest patches, ranging from a few hectares in size to over 8 hectares, lying next to rice paddies are also valuable in terms of bird habitat conservation, although mostly for common species. There are several forest patches close to the routes, which Tsukuba City government has attempted to protect from deforestation, promoting satoyama activities concurrently. One of them provided breeding habitat for a pair of buzzards every year during the study period. The Japanese city planning act controlled urbanization, where rice paddies remain as long as the landowners are willing to maintain. Preservation of secondary forests would be the most challenging issue for avian conservation as well as for the provision of comfortable environments for visitors.

Many flowers bloom along roads, and visitors can enjoy a variability of flowers during different seasons. I was interested in whether the herbaceous flowers attracted them, and counted the number of species four times a year, supposing that each period represented a 3-month season. I cycled and looked for flowers along the same routes as I used for monitoring birds and visitors. It was a crude estimate of species richness, and I did not endeavour to find all the species on the routes. Some flowers were so small that I surely have overlooked some of them. Some species were similar to another species of the same genus, so sometimes I identified the genus but not the species. Yet such a crude method was able to find a large degree of variance in the number of species among the seasons (Table 5.5).

The number of visitors did not decline in winter when there were a small number of flowering species. I also measured temperature (9.1 degree Celsius in January on the average and 28.2 degree Celsius in August) without finding any correlation with the number of visitors. More interesting was that there were many indigenous species despite such habitat conditions disturbed by exotic plant invasions, occasional herbicide spray and pesticide transported from farmland nearby. Comparing herbaceous species with Amami and Tadami is out of the scope of this book, but it is evident that their species diversity was high in Tsukuba. Tsukuba has distinctive four seasons, while Amami has a more extended summer and Tadami has a longer winter than Tsukuba. Besides, a large area of farmlands provides many more patches of open habitat for herbaceous plants than the forested areas in Amami and Tadami, contributing to species richness. In contrast, I was able to find only one species endemic to Japan, *Chrysosplenium grayanum*. I do not give species lists for Amami and Tadami, but there are many more endemic species.

Dividing the routes into the same sections with the census of birds and visitors, I evaluated species richness in terms of the number of species per section. I was not able to obtain a significant correlation between flowering species richness and the number of visitors. I think the visitors were more concerned with landscape with different types of ecosystems than small plants flowering at their feet.

Table 5.5 Number of flowering herbaceous species and families along the survey routes in satochi

Family/Survey season	*January*	*April*	*July*	*October*
Acanthaceae			1 (1)	1 (1)
Alismataceae			1 (1)	
Apiaceae			1 (1)	1 (1)
Asphodelaceae			1 (1)	
Asteraceae	5 (3)	9 (6)	13 (6)	15 (6)
Boraginaceae		1 (1)		
Brassicaceae	2 (2)	5 (4)	3 (3)	4 (4)
Campanulaceae			1 (1)	
Caprifoliaceae			1 (1)	
Caryophyllaceae	2 (0)	2 (0)	1 (0)	2 (1)
Commelinaceae			1 (1)	1 (1)
Convolvulaceae			1 (1)	2 (1)
Crassulaceae		1 (1)	1 (1)	1 (1)
Fabaceae		3 (1)	3 (1)	3 (1)
Gentianaceae		1 (1)		
Geraniaceae		1 (1)		1 (1)
Iridaceae			1 (0)	
Lamiaceae	2 (1)	3 (2)	5 (4)	1 (1)
Liliaceae		1 (1)		
Linderniaceae			1 (1)	
Mazaceae		1 (1)		
Onagraceae			2 (1)	2 (1)
Oxalidaceae		1 (0)	2 (0)	2 (0)
Papaveraceae		2 (1)	1 (1)	
Phytolaccaceae			1 (0)	1 (0)
Plantaginaceae		1 (0)		
Polygonaceae		1 (1)	1 (1)	3 (3)
Primulaceae			1 (1)	
Ranunculaceae			1 (1)	1 (1)
Rosaceae		2 (2)	1 (1)	
Rubiaceae			1 (1)	
Saururaceae			1 (1)	
Saxifragaceae		1 (1)		
Scrophulariaceae	1 (0)	2 (0)	1 (0)	3 (2)
Solanaceae			2 (1?)	2 (1?)
Tiliaceae				1 (1)
Violaceae		1 (1)		
Vitaceae			1 (1)	1 (1)
Total	12 (6)	39 (25)	52 (34?)	48 (28)

The number of indigenous species is in parentheses.

A species of Soanaceae was difficult to distinguish between native and exotic species.

For the classification of the family to which the species belongs to, the website provided by the Weed Science Society of Japan (5-8) is referred to.

Special thanks to M. Ishizuka for his assistance in species identification.

5.4 Values of satochi-satoyama and mountains near large cities

This chapter suggests that provision of leisure space by satochi and satoyama should be emphasised rather than biodiversity at the national level. As far as forest biodiversity conservation is concerned, we should give more emphasis on old forests in Amami and Tadami than young forests in satoyama. Biodiversity-rich satoyama is composed of various landscape components, while coppice forests and traditional landscape management have been emphasised in the literature repeatedly. However, the area of such forests is minimal and fragmented as indicated before. Conventional agricultural management practices have almost disappeared. Therefore, their arguments are not compelling.

People use satochi frequently as long as the landscape is sufficiently attractive and because it is close to residential settlements. Several websites mention that Mt. Takao (599 m high) is the most crowed mountain in the world. I am not sure about it, but over 3 million people were estimated to have visited the quasi-national park of total area 770 hectares in 2016 according to the statistics given by the Ministry of the Environment[(5-9)]. One of the reasons why it is so crowded is easy access from Tokyo. For instance, it takes only a few hours from one of the busiest train stations in Tokyo. Mt Tsukuba area is also crowded, and Tsukuba City Government estimated that 3.2–3.8 million people visited annually during the decade 2007–2016[(5-10)]. Iwata (2015) indicated that about 50% of the visitors were from the prefecture to which Mt Tsukuba belonged and 97% were from areas within a distance of 80 km. It is less than 1 h by car from many cities east of Tokyo, and it is within a few hours from Tokyo Central Railway Station. In contrast, a smaller number of people, 1.2 million, visited one of the most famous tourist spots in the country, Kamikochi, than the two mountain areas near Tokyo in 2016[(5-11)]. Kamikochi has majestic natural beauty located in the Northern Japan Alps, about 200 km away from Tokyo. These results suggest that access and population size combined would have a greater effect on the number of visitors than the quality of natural beauty.

About one-fourth of the Japanese population is in the Tokyo metropolitan area. Many of them cannot help getting out of small living spaces and enjoying natural environments. They do not have to climb up Mt Tsukuba and Mt Takao on their efforts, but a ropeway or a mountain tram carries them to the top. Such locations are minimal near Tokyo and a few mountains have such conditions.

Supplementary notes

5-1) satochi landscape is dominated by farmlands, whereas woodland is dominant in satoyama. Both of them have fundamentally similar landscape components. In other words, satochi is a term used by agronomists, while satoyama is the forester's view of a landscape.

5-2) *Nihon no satochi satoyama no chōsa bunseki nitsuite: chūkan hokoku* (Kankyō-shō). Available at: https://www.env.go.jp/nature/satoyama/chukan.html (accessed 22.01.20).

5-3) Information in this paragraph is written in a dissertation by Mizuki Kawano for her bachelor degree from Nagasaki University.

5-4) Based on the statistics on the wooded areas provided by the city government. Available at: https://www.city.tsukuba.lg.jp/kurashi/index.html. (accessed 3.02.20).

5-5) The government made maps called Jinsoku sokuzu in the late nineteenth century, which drew planted tree species as well as other geographical features. Available at: https://habs.dc.affrc.go.jp/habs_map.html?zoom=13&lat=36.08382&lon=140.19843&layers=B0. (accessed 10.04.20).

5-6) Mt Tsukuba has two peaks of almost equal height. There is a quasi-national park that has special and first-class protection zones around the peaks. These zones also cover the southern slopes down to 300 m, where evergreen broadleaf trees cover more extensively than the other side.

5-7) *Nihon no gairaishu lisuto, Kokuritsu Kankyō Kenkyūsho.* Available at: https://www.nies.go.jp/biodiversity/invasive/index_en.html (accessed 5.02.20).

5-8) *Zasshu-lisuto.* Available at: http://www.wssj.jp/~term/weed_name_list.html (accessed 3.02.20).

5-9) *Shizen-kōentō riyōsha sūchō* (Kankyō-shō). Available at: https://www.e-stat.go.jp/stat-search/files?page=1&toukei=00650103&tstat=000001030999 (accessed 12.04.20).

5-10) *Dai-ni-ji Tsukuba-shi kanko kihon keikaku.* Available at: https://www.city.tsukuba.lg.jp/kankobunka/kankojoho/1001464.html (accessed 12.04.20).

5-11) *Kankōchi riyōsha tōkei chōsa* (Nagano-ken). Available at: https://www.pref.nagano.lg.jp/kankoki/sangyo/kanko/toukei/riyousya.html (accessed 12.04.20).

6 Regional comparison and summary discussion

6.1 Wildlife and forest uses in northern Kyushu in comparison with other regions

In northwestern Kyushu, plantations dominate forest landscapes. The region extends more than 130 km from Nagasaki in the west to Hita in the east (Figure 6.1). Mountains rise to peaks of 1231 m in Hita, 1055 m in Saga and 1483 m in Nagasaki. This region is the fourth case study area (Figure 1.2). The most significant difference in landscape composition from the other regions is that commercial trees—sugi *Cryptomeria japonica* and hinoki *Chamaecyparis obtusa*—dominate the forested area: not because sugi is more productive than in the other regions, but because it grows in most areas. Productivity measured as trunk volume per hectare is slightly lower in Kyushu than in Tohoku until stand age 40 years and is notably lower at age 60 years (Table 6.1). However, the area of plantations in Tohoku is limited owing to heavy snowfall, as indicated in Chapter 4. The Nansei Archipelago is too warm for commercial timber production, and soil conditions are unsuitable. Forestry is a minor land use in the satochi landscapes in the Kanto region. Other differences are that there are not many terrestrial endangered species in the forests in northwestern Kyushu and that local people are less interested in biodiversity conservation.

This section relies primarily on statistics on endangered and endemic species, land use and tourism, and on personal observations, as there are few studies related to these topics.

6.1.1 *Wildlife and forest uses in landscapes dominated by plantations*

Jointly, Nagasaki, Saga, Fukuoka and Oita prefectures average 53.7% plantations by forested area, and all exceed the national average of 40.7%[(6-1)]. Saga, a prefecture located east of Nagasaki, is proud of its percentage (67.3%), ranked as the highest in the country. Plantations cover 76% of the total forested area (207 km^2) of Hita City at the western edge of Oita Prefecture (Nōrin-suisan-shō tōkeijōhō-bu, 2013). In northern Kyushu, two mountain ranges along a tributary of the Chikugo River, separated by a small valley, isolate the forests of Saga

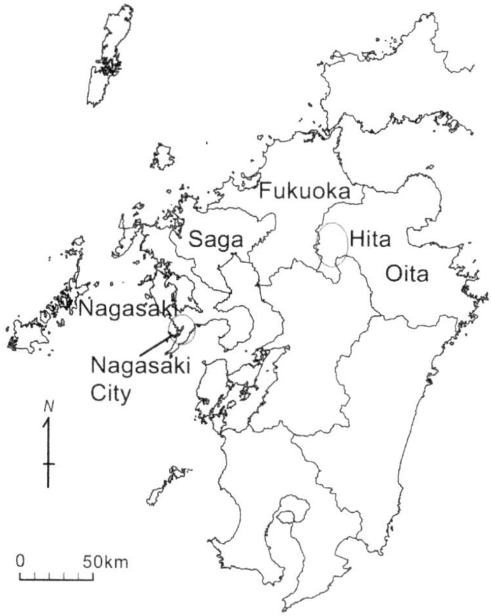

Figure 6.1 Locations of four prefectures in northern Kyushu and Hita and Nagasaki City.

and Nagasaki from the central mountains of Kyushu, which extend further east and south of Hita. There is a national park in the centre of Oita Prefecture that has different landscapes from those dominated by tree plantations.

The study area spans from the western part of Oita westward. Within this range, the forests from Hita to western Saga have one of the highest proportions of plantations in Japan. In contrast, the poor soils in Nagasaki Prefecture explain the lowest proportion of plantation forests among the four prefectures, at 43%. The other prefectures have fertile volcanic soil suitable for growing sugi, which grows faster and is more economically competitive than hinoki. Increasing demand for timber in China has motivated loggers to cut large areas of sugi.

Table 6.1 Comparison of trunk volume of sugi (m^3) per hectare between Tohoku and Kyushu region

Region/Stand age	*20 yr*	*40 yr*	*60 yr*
Kyushu (9)	186	477	665
Tohoku (52)	198	485	741

Stand volumes were measured by Shinrin sōgō-kenkyōsho (National Institute of Forest Sciences) in various plots established to monitor long-term trends in trunk growth. The number of plots is in parentheses. The tree trunk volume is the average of the estimated volumes based on the measurements in each plot. I collected the data by myself when I was working in the institute.

Forest managers in Nagasaki Prefecture planted mostly hinoki on the mainland and sugi on the islands. Plantation forestry is not very profitable and many forests are next to urbanised areas, but they have little interests in public utility of forests. When asked about the functions other than wood production in a prefecture committee on forest management, the government official answered that they must resolve silvicultural problems first, indicating that they had little interest in public utility, as different challenges continue to crop up.

Japan is home to three large mammal species that are abundant and considered pests of agriculture and forestry, namely deer *Cervus nippon*, wild boar *Sus scrofa* and monkey *Macaca fuscata*. Others not regarded as pests are serow *Capricornis crispus* and bear *Ursus thibetanus*. All have a wide distribution range in Honshu, from the Tohoku to the Kinki regions, and not threatened with extinction. Field signs of serows—faecal pellets, footprints and hair—were often visible in forests of Tadami. Bear hunting used to be popular before the power plant accident, as mentioned before. Further to the west, however, bear populations are small and fragmented, and serows do not live in the westernmost part of Honshu (Kankyō-shō, 2004). In Kyushu, in contrast, the serow may be threatened with extinction (Sakata *et al.*, 2013), having a small, isolated distribution range in high mountains about 30 km east of Hita (Yasuda *et al.*, 2013). The bear was considered extinct in Kyushu during the middle of the last century (Ohnishi and Yasukochi, 2010). In addition, two arboreal species of mammals, still common in Honshu, have gone extinct or are on the brink of extinction in Kyushu (Yasuda, 2007). One of them, squirrel (*Sciurus lis*) has not been reported since the 1990s and the other, the dwarf flying squirrel (*Pteromys momonga*) has small and fragmented populations in the high mountains (Yasuda, Ibid). Yasuda (Ibid) suggested that the small proportion of natural forests in Kyushu is one reason why those two species are not found or are scarce. This reason may also explain the small number of serow and the absence of bear. No mammals in northern Kyushu are on the national Red List[6-2].

Unlike on Amami and in Tadami and Tsukuba, no local volunteer groups in northwestern Kyushu conducted intensive bird surveys. According to two sources of crude nationwide information[6-3], northwestern Kyushu has a total of 109 bird species in landscapes dominated by forest, including six species of birds on the national Red List: Indian pitta *Pitta brachyura* (EN), grey-faced buzzard *Butastur indicus* (VU), *Pericrocotus divaricatus* (VU), ashy minivet *Eurystomusu orientalis* (VU), western osprey *Pandion haliaetus* (NT) and Eurasian sparrowhawk *Accipiter nisus* (NT). Northwestern Kyushu has fewer endangered mammals and birds than Amami and Tadami (Table 6.2), although covers a much larger area, and a similar number to that in Tsukuba, which has a comparable 119 bird species (Chapter 5) and number of endangered species. Among the six Red List species in northwestern Kyushu, two species, *Pitta brachyura* and *Eurystomusu orientalis* were not found in Tsukuba, as they are spring migrants whose distribution does not extend to eastern Japan, and live in low-altitude broadleaf forests (Ueno, 2009; Fukui *et al.*, 2005). The reason for their absence is geographical, not the difference in habitat conditions.

Table 6.2 Number of endangered wildlife species by national Red List category

	Mammals				*Birds*			
	CR	*EN*	*VU*	*NT*	*CR*	*EN*	*VU*	*NT*
Amami	1	6	0	1	0	0	4[a]	0
Tadami	0	0	2	0	1	3	5	5
Tsukuba	0	0	0	0	0	0	3	2
Northern Kyushu	0	0	0	0	0	1	3	2

Note:

a One woodpecker species (see Table 2.1) is listed at subspecies level. All the other species in the table are species level.

Due to the Ministry of the Environment policy, information on birds of prey is not included in the table as it is not disclosed to the public or for scientific research purposes.

The one endangered species found in Tsukuba but not in northwest Kyushu is the plumed egret *Egretta intermedia*, a summer migrant in Tsukuba. The reason for its absence in northwest Kyushu is most probably relatively small areas of rice paddies: in other words, satoyama landscape. Thus, the number of endangered species per area is much lower in northwestern Kyushu than in Tsukuba.

Northwestern Kyushu and Tsukuba are comparable in having no nationally endangered species of mammals and no bears or serows, and in having an endangered salamander species. The salamander in northwest Kyushu (*Hynobius nebulosus*) lives at low elevation and is ranked lower as VU than the other species in Tsukuba, which lives at higher elevation and is ranked as CR. However, despite such similarities, given the size of northwestern Kyushu, species diversity per area there is low. There are no records that show that the salamander in northwest Kyushu depends on highland habitats, and thus it is considered a satoyama species, probably because plantations dominate at altitudes above satoyama. Also, the dominance of plantations may be one of the reasons why squirrels are common in Tsukuba (Ibaraki-ken Shizen-hakubutsukan, 1998) but are not found in northwestern Kyushu. As described in Chapter 5, some species are found only or more frequently at higher elevations in the Mt Tsukuba area, where natural forests are protected. Another difference is that the forest area in Tsukuba is much smaller than in northwestern Kyushu, so deer and monkeys are absent, and populations of wild boar are not high.

To compare bird species richness between plantations and natural forests, I counted birds that inhabit plantations at 23 points in Hita. The method was the same as that used on Amami (Chapter 2), and in Nagasaki and Tadami. Hita is famous for producing high quality timber, and plantations of sugi and hinoki cover more than 90% of the land in uninhabited areas (Photo 6.1). The well managed plantations had no broadleaf trees in the canopy layer, but broadleaf trees grew in lower layers under timber trees and small patches along the roads around the census plots. There was only a 30 hectares of forest reserve set aside

Photo 6.1 Hita's okuyama landscape predominated by tree plantations that extend as far as the eye could see.

to protect *Fraxinus spaethiana*, a species of ash endemic to Japan, on top of a mountain. This species is not so rare to be included in the national Red List. The average numbers of species and birds per census plot were less in Hita plantations than in natural forests (Table 6.3). The total number of non-migratory species in Hita was also lower than that on Amami and in Tsukuba. Similar results were obtained when comparing natural forests and plantations in Tsukuba, although the sample size was too small to reach a definite conclusion. The survey was

Table 6.3 Comparison of the number of forest bird species and abundance observed per census plot between study areas; average and SD (in parentheses)

Area (Number of plots)	*Species per plot*	*Number of birds per plot*	*Total number of species*	*Altitude range (m)*
Amami (51)	6.2 (1.4)	9.1 (2.4)	26	175–455
Tadami (22)	4.7 (1.5)	7.0 (2.9)	26	450–830
Tsukuba (10)	5.1 (1.8)	6.4 (2.4)	34	190–835
Nagasaki (26)	4.8 (1.0)	8.3 (2.2)	13	120–520
Hita (23)	2.1 (1.3)	3.2 (2.5)	14	340–870

The survey period was April to May 2017, June 2019, May to June 2006 and 2007, April 2019 and March 2019 for Amami, Tadami, Tsukuba, Nagasaki and Hita, respectively.

conducted in Hita in March, before the arrival of spring migrants, and in the other areas from April to June. In Nagasaki City, spring migrants stop by, but according to my five years of observation, they do not breed there. Even if a small number of migrant species breed in Hita, the presence of migrants would not significantly affect the results in Table 6.3. Hita city's environmental policy has no description of biodiversity conservation strategies, suggesting that locals consider forests to be a timber resource and not a place for biodiversity conservation.

The forest slopes adjacent to the urban areas of Nagasaki City have few plantations, a sight rare in northwestern Kyushu. The city is densely populated under the mountain slopes whose locations are not suitable for planting commercial trees because clear-cutting must increase the risk of slope failure. Forests there have clear evidence of past coppice management. Photo 6.2 shows multiple stems sprouting from tree bases, suggesting that the forest was left unmanaged after logging for the production of firewood and charcoal. Almost all the canopy species are evergreen broadleaf trees such as *Quercus acuta*, *Quercus glauca*, *Machilus thunbergii* and *Cinnamomum Camphora*. The bird count studies suggest that the numbers of species in these forests were as low as those in Hita (Table 6.3). One reason would be that these forests are geographically isolated (Figure 6.1) from the more species-rich central mountains in Kyushu. Note that Saga Prefecture east of Nagasaki has the highest percentage of tree plantations in the country. Another reason could be that the forests have no old-growth remnants.

Photo 6.2 Coppice woodland near urban areas of Nagasaki City that have been neglected for more than 50 years.

On the other hand, the numbers of species and birds per census plot were much higher in Nagasaki than in Hita. These results indicate that tree plantations are species-poor as far as canopy trees and birds are concerned.

6.1.2 *Satoyama activities and nature conservation in northwestern Kyushu*

When forest landscapes in northwestern Kyushu are compared with the other case study areas, the most prominent feature is the high proportion of plantations in okuyama, in uninhabited regions. On the other hand, the fundamental composition of the satoyama landscapes are comparable except on Amami, which has few paddy fields. The government has encouraged sugarcane production through subsidies on Amami, and only a small community grows rice to maintain the traditional culture. There is no satoyama activity group on Amami. Broadleaf woods used to be coppiced, but locals are not interested in restoring such forest management.

In other regions, local people are involved in various kinds of satoyama activities similar to those described in the last chapter (Table 5.1). There is no significant difference in the number of groups per population unit between the Kanto and Kyushu regions (Table 6.4). Kawano (unpublished) examined the activity programs publicized on the websites, and found a clear difference in the forest types of activity sites between Kanagawa and Nagasaki prefectures: 91% of forest management activities were practised in broadleaf forests in Kanagawa, versus 44% in Nagasaki. The rates of activity in broadleaf forests in other prefectures also showed a clear difference: 15% in Saga (Kyushu) versus 74% in Ibaraki (Kanto). Conversely, the percentage of groups working in plantations and bamboo groves is higher in northwestern Kyushu than in the Kanto region.

One factor responsible for such differences may be that the most popular tree in the Kanto region, *Quercus serrata*, does not grow in the lowlands of Kyushu where evergreen trees are predominant. Deciduous trees keep the forest floor brighter and provide a more comfortable working environment than in dark

Table 6.4 Comparison of the number of satoyama activity groups with respect to population size in three regions

Region	*Number of groups*	*Population (million)*	*Number of groups per million people*
Tohoku	107	9.81	10.9 (3.8)
Kanto	302	40.69	7.4 (1.9)
Kyushu	96	13.45	7.1 (3.4)

The data for satoyama activity groups was reported in 2001 and population was from the same year.

SDs among prefectures in each region are in parentheses.

Source: Kankyō-shō for the number of satoyama activity groups[6-4] and Sōmu-shō for population[6-5].

evergreen forests. The most dominant canopy species in naturally regenerated forests in Tsukuba is the evergreen *Quercus myrsinaefolia*, but locals prefer to plant *Q. serrata* for satoyama activities. Another factor may be that in northwest Kyushu, forestry experts provide advice and sometimes play a leading role based on their silvicultural knowledge and skills.

Conservation is a frequent type of citizen activity. On Amami, several local conservation groups have fought to protect endangered species from activities such as golf course development, chip mill construction and the construction of a large dam for water storage. Their concern is not forest management but the conservation of endangered species habitats. They are so active that local governments cannot ignore them in large-scale developments. In Tadami, people are more interested in nature conservation than in satoyama activities. The town office runs a small museum that displays flora and fauna of the beech forest ecosystems, and explains the state of forest use (described in Chapter 4). Local concerns are natural forests and the use of plants and animals for food and recreation. Regional differences are evident. Conservation activities are frequent in areas with many nationally endangered endemic species. In contrast, satoyama activities are conducted in areas with many people living relatively close to forests.

Regional differences can also be seen in recreational activities. A good example is harvesting wild plants and mushrooms, which is more common in areas with significant snow accumulation each year (Kawarasaki and Sugimura, 2012). It is not faddish to collect non-timber forest products (NTFPs) in Nagasaki except for spring bamboo shoots. In this context, a group of ethnologists maintain that the traditional culture of southwestern Japan derives from the Yunnan region of China, where evergreen forests predominate (e.g., Ueyama, 1969). The people burned and felled forests, planting rice, potatoes, vegetables and small trees, but did not make much use of NTFPs. Such forest uses are in contrast to those in northern Honshu, where many plant and animal species are used for food and other purposes without significantly modifying the forest ecosystems (Kitamura, 1998).

When the top tourist destinations are compared, the difference in popularity of nature-based tourism between northwest Kyushu and Kanto is evident (Table 6.5). Natural areas feature terrestrial forested landscapes, hence excluding coastal areas, artificial facilities such as urban parks, buildings in mountain areas and notable monuments such as huge trees or spectacular rocks. Table 6.5 shows the number of popular natural areas of the top 50 sites ranked by website content and suggests two differences between the two regions. First, there are more types of tourism and destinations in the Kanto region than in Kyushu. Second, all prefectures in Kanto have more tourist destinations than Kyushu except for Oita Prefecture. Oita has the highest number of popular sites in northern Kyushu, but all the sites are located in the national park east of Hita. Nagasaki Prefecture has several mountains with well-managed paths to the summits, but none of them is ranked very high. The natural area of Nagasaki Prefecture is larger than that of four Kanto prefectures, and that of Oita is larger than that of

Table 6.5 Comparison of popular nature-based tourism destinations between northern Kyushu (upper four rows) and Kanto region (lower six rows)

Prefecture	*Tourism type; destination*	*Number of popular sites*[a]	*Natural areas (1000 hectare)*[b]
Nagasaki	VM; HL	3	135
Saga	HM; RV	4	30
Fukuoka	HM; HL, RV	3	63
Oita	HM, DF; VM, LK, RV	6	191
Tochigi	HM, DF; HL, LK, RV,	17	184
Gunma	HM, DF; HL, LK, RV	13	218
Saitama	HM, VM; LK, RV, FP	12	61
Ibaraki	HM, VM; LK, RV	9	67
Chiba	HM, VM; LK, RV, FP	7	91
Kanagawa	VM; HL, FP	6	55
Fukushima	HM, DF; LK, RV	10	944

Tourism type: HM, hiking in mountains; VM, view from the mountain top; DF, driving through forested areas; HL, highland; LK, lakes; RV, rivers; FP, forest park. Refer to the text for the selection criteria.

Note:

a The website(6-6) reports rankings every month. Winter months, December to February, are excluded.

b The areas include forests and grasslands under the jurisdiction of the Forestry Agency.

Source: Jaran Kanko spot ranking(6-6).

five Kanto prefectures (Table 6.5). Yet these sizable natural areas do not translate to a high number of popular sites in either prefecture, suggesting the significance of the quality of the natural environment over quantity. Another factor that may have caused these differences between Kanto and northern Kyushu is people's innate preference for nature, which is seen also in satoyama activities. Given the size of the population, we see that nature-based tourism is much more prevalent in Kanto than in northern Kyushu. Fukushima Prefecture has more popular tourism destinations than northern Kyushu prefectures, but this preference may be influenced by the size of the natural areas.

A comparison of national parks in the northwestern part of Kyushu with other regions reveals its characteristics. There are two national parks in Nagasaki Prefecture but none in Saga and Fukuoka Prefecture. One of them, Unzen Amakusa National Park (NP), on the peninsula south of Nagasaki City, is one of the eight oldest national parks established in 1934. The primary purpose of the original national parks, under the jurisdiction of the Ministry of Health until 1971, was to protect the beautiful scenery for travellers. At the same time, the central government expected to promote the local economy by attracting tourists. The most strictly protected zones—special protection and first-class protection combined—account for 5.6% of the park's area. These zones are much smaller than those in the most recently established parks, Amami Gunto NP

(34%) and Oze NP (42%) in southwestern Fukushima. The other national park in Nagasaki is Saikai NP, and its most distinctive landscape feature is a mixture of intricate coastlines and small islands. There are no endangered vertebrates other than the salamander, which lives in satoyama outside the most rigorously protected zones of the park. Sasebo, the prefecture's second-largest city, has allowed the conversion of forest in which the salamander had maintained a small population to a residential area. This action represents the poor state of biodiversity conservation in Nagasaki, maybe because there are no symbolic species that promote conservation activities, such as the Amami rabbit and the golden eagle in Fukushima.

The local societies in northwest Kyushu are not very interested in biodiversity, but damages to agriculture by the wild boar (*Sus scrofa*) and the Japanese deer (*Cervus Nippon*) are matters of considerable concern. Table 6.6 compares the damages by three species in four prefectures in northern Kyushu and four in eastern Japan. Wild boars and deer do more damage on average in northern Kyushu than in eastern Japan, particularly per unit area of farmland. The affected area can be an indicator of a species' population size. The three species live in forests, and the boar population density appears to be high in northern Kyushu. Deer are less damaging than boar, but some dense populations have almost wholly removed forest floor vegetation. On the other hand, in Fukushima Prefecture, the damage caused by deer is small. The deer populations in the northern Kanto region have not moved to the eastern part of Fukushima Prefecture. Heavy snow prevents the boar from becoming dense in western Fukushima, so damage there is much smaller than in northern Kyushu. In contrast, monkeys are well suited to withstanding winters with heavy snow cover, maintaining relatively large populations in eastern Japan.

Table 6.6 Damages to agriculture by wildlife species

			Damaged area (hectare)			*Monetary amount (million yen)*		
Prefecture	*Farmland (thousand hectare)*	*Forest (thousand hectare)*	*Boar*	*Deer*	*Monkey*	*Boar*	*Deer*	*Monkey*
Nagasaki	54	248	187	10	0	142	8	0
Saga	59	110	79	0	1	92	0	5
Fukuoka	95	223	209	57	1	307	60	3
Oita	65	457	88	68	3	99	24	9
Average	68	260	141	34	1.3	160	23	4
Fukushima	166	944	92	14	10	98	4	18
Ibaraki	183	190	86	0	0	100	0	0
Tochigi	133	345	120	40	10	128	29	19
Kanagawa	22	95	24	5	3	38	32	49
Average	126	394	81	15	5.8	91	16	22

Farmers voluntarily submit estimated amounts of damage to the local government office. After collecting information from all municipalities, each prefecture reports the sum of the costs to the Ministry of Agriculture.

Source: Nōrinsuisan-shō(6-7).

In summary, northwestern Kyushu has fewer endangered terrestrial vertebrates and active conservation groups than Amami and southwestern Fukushima. Satoyama activities tend more toward forestry than in the Kanto region and Fukushima. Nature tourism in the mountains is less favoured in northern Kyushu than Kanto and Fukushima. The national parks in Nagasaki have much smaller proportions of most strictly protected zones than the most recently established parks on Amami and in Fukushima. These results, coupled with the high proportion of plantations, suggest that the local governments recognise timber production as the main role of forests in northwest Kyushu and is less concerned with biodiversity than in the other regions.

6.2 Satoyama in West Java

The following case study in West Java, Indonesia, serves as an excellent reference for finding a better balance between forest management and wildlife conservation. One reason is that the landscapes of satoyama in Indonesia and Japan are very similar. The study area in Java has many rural settlements surrounded by rice paddies, dry fields, orchards and tea gardens, with elaborate irrigation systems, at the foot of mountains whose peaks rise above 2000 m. The main differences are that Japanese satoyama generally has a larger area of tree plantations and that Javanese satoyama is utilised much more heavily in more ways for subsistence living. The second reason is that the West Java forests retain high levels of species diversity despite heavy use by locals. The third reason is that people live in national parks, as in Japan. However, Indonesia's national park system has different characteristics from the Japanese system, so it may provide an appropriate reference to reveal the relative strengths and weaknesses of the Japanese system.

There are some common features of forests and biodiversity conservation in Indonesia and Japan. Many residential areas have high population density near forests in both countries. Many local communities depend on forest resources in their daily lives, and forests have been managed through participatory processes by communities. Besides, many genera and species are common to both countries, and therefore biogeographical boundaries are drawn so that the southernmost islands of Japan and Java belong to the same region, Oriental. This section presents two studies in Gunung Halimun Salak National Park (GHSNP), where I worked with research assistants, Age Kridalaksana and Wim Ikbal Nursal for about two years at the Center for International Forestry Research.

The idea of national parks that originated in the United States spread to Europe and was brought to Indonesia by the Dutch colonies (Jepson and Whittaker 2002; Usuki 2004). The parks were established to maintain pristine natural ecosystems. Mostly they consist of strictly protected zones, where no one is allowed to use forest resources and ecosystems, but recreation, education and research are permitted. The Ministry of Forestry (now Ministry of Environment and Forestry), which owns and manages the national parks

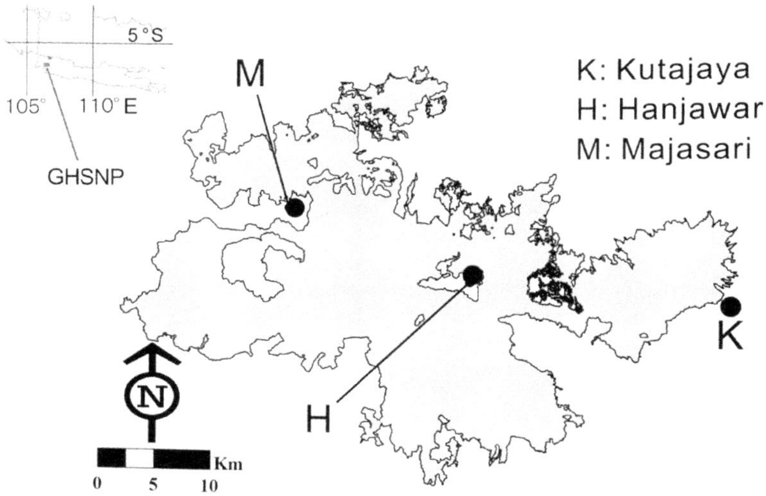

Figure 6.2 Location of GHSNP, Java, and three residential settlements.

attempted to drive out all the people who lived there. Following struggles between the government and locals in and around the parks. So the government changed its policy in 2006, allowing locals to use parts of forests in the parks. Instead of moving people out, the GHSNP office developed residential enclaves, within the park (Figure 6.2), experimenting with the most advanced ways of park management in Indonesia.

The Indonesian national park zoning system (Table 6.7) is more complicated than the Japanese system. In the Indonesian system, the core, wilderness and utilisation zones, correspond with IUCN categories I and II, and the rehabilitation zone is designated for restoration. The special zone seems to be similar to IUCN category V and Japan's second-class special zone. The other two zones, traditional and religious use, complicate the overall system and could be considered parts of the special zone. To further simplify the system, areas of the rehabilitation zone may be integrated into either the core or wilderness zone in the future. The Japanese system seems to have a workable structure in which national parks include extensive private lands (Chapter 1).

GHSNP and its surrounding areas have a long history of agricultural development. Parts of it were deforested through expansion of farmlands because many people depended on paddy cultivation and other agriculture in and around the park. The most critical challenges for the park management are to prevent the expansion of agricultural lands due to population growth, to preserve biodiversity and at the same time improve local livelihoods. Indonesian national park management prohibits the harvesting of natural resources within park boundaries, except in limited zones, mostly the special zone. Locals harvest forest products such as firewood, edible plants, bamboo, rattan, sugar from sugar palm and grasses on the forest edges (Sugimura *et al.*, 2015). GHSNP has 314 settlements.

Table 6.7 Indonesian national park zoning system

Zone	*Key features*	*Role*	*Allowed activities*
Core	Pristine, genuine and unique ecosystems	Natural laboratory, germplasm source, sustaining natural ecological process	Scientific research, conservation education, non-permanent supporting facilities
Wilderness	Wildlife home range and habitat	Lying around core zone to mitigate impacts from other zone types	Scientific research conservation education, restricted tourism, supporting facilities
utilisation	Unique, attractive and beautiful ecosystems and geologic features	Natural attraction	Tourism and tourism development, ecosystem services use, supporting facilities
Rehabilitation	Degraded or deforested land, exotic species eradication site	Rehabilitation site, habitat expansion	Rehabilitation related activities
Traditional	Degraded forests with long history of traditional use by local community	Natural resources for traditional community livelihood	Sustainable use of natural/forest resources under regulation
Religious use, protection of culture and history	Historical and cultural sites and religious activity sites	Preservation of historical and cultural sites	Ritual exhibition, cultural/historical sites maintenance
Special	Location with group of people and existing infrastructure prior to park designation	Space for daily activities of group of people	Rehabilitation related activities, resource utilisation for livelihood

About 100,000 people depend on the natural resources in the park, and most people are self-sufficient (Kubo and Supriyanto, 2010), collecting forest products daily and cultivating arable land. However, the primary objective of the national parks is to protect the natural ecosystems, while minimising human influences. At the time of our research, the core and wilderness zones (Table 6.7) combined accounted for 51% of the area and the rehabilitation zone for 27%, but the other zones were still in review by the park office. The government allocated relatively large proportions of areas to these strictly protected zones, despite the need of local for special zones.

6.2.1 *Utilisation of forest resources by local people*

We conducted surveys in and around three residential settlements in GHSNP in 2012; Kutajaya, Hanjawar and Majasari. Hanjawar lies within the national park boundary (the shaded portion of the map in Figure 6.2). Majasari lies outside the park but is nearly surrounded by it. The residents of Hanjawar and Majasari relied mainly on agriculture and depended more on NTFPs than

Kutajaya, which lies outside the park (Ibid). The three settlements stood out from each other. Many inhabitants of Kutajaya worked in *Agathis damara* plantations that were incorporated into the park in 2003. During the last years of the Dutch colonial period, tea plantation workers built the Hanjawar, which accessible only by motorcycle. Majasari has the most complicated history. It was built by people who worked as labourers in the Dutch quinine plantations when the land was not allowed to be cultivated. The Japanese army then converted forests into fields and gardens to meet military food needs, while some people were allowed to own parcels of land that are now within the park. Following Indonesia's independence, the forests were managed by the Ministry of Forestry as sanctuary or production forests. The ministry did not permit the conversion of forests to farmland but failed to control cultivation owing to lack of staff. In 2003, GHSNP incorporated surrounding areas, including production forests and cultivated lands. Heated arguments and confrontation between the residents and the park office continued until 2007, when the residential areas were authorized at a meeting between GHSNP officers and locals.

In our study, we asked ten local people in each village to help us collect information on NTFP harvesting and use and to carry a Global Positioning System (GPS) receiver (Trip Recorder 747Pro) for seven days. We conducted the survey in cooperation with a Bogor-based organization, Japan Environmental Education Forum, and students from Bogor Agricultural University. We identified the location, local name, scientific name, parts of plant harvested, purpose and frequency of collection and amounts of each product harvested. We classified the species into eight groups according to their use:

(A) Firewood: plants used as firewood and for bundling firewood
(B) Fodder for cattle
(C) Sugar collection: sugar palm trees and plants used for sugar processing
(D) Medicinal use: plants used as traditional medicine
(E) Food
(F) Construction materials: plants used to build houses
(G) Non-food foliage: plant leaves used to wrap food or other objects
(H) Household appliances

There were obvious differences in the frequency of NTFP collection among the three settlements (Figure 6.3). The villagers harvested fodder plants most frequently in Kutajaya and Hanjawar where they kept goats as "savings" for when large sums of money were needed. They fed livestock almost every day, using one or two bundles weighing 25–50 kg each. Firewood was frequently collected in all the settlements, and the total weight gathered was not significantly different: About one to two bunches per day, equivalent to 25–50 kg. All respondents used it for home cooking. Kutajaya villagers did not spend much time collecting anything other than firewood and fodder, possibly because most

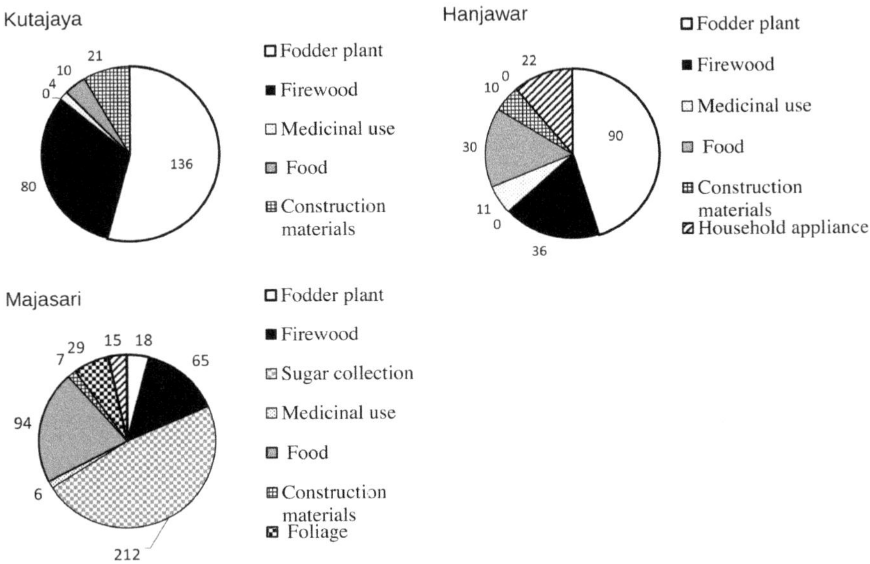

Figure 6.3 Frequency of NTFP harvesting by ten people in each settlement in seven days.

of them worked in a few non-agricultural industries nearby or perhaps because the *Agathis* plantations they visited did not have many species for NTFP use. Unique among the three settlements was that Majasari villagers collected palm sugar and sold it to their neighbours or a middleman in the village. Although sugar palms grew in Hanjawar, they were not used probably owing to inconvenient transportation and access to townships. There were no sugar palms in Kutajaya.

For non-food items, such as firewood, fodder plants and foliage, use is biomass based and species are less important than quantity. For the other groups, in contrast, use is biodiversity-based, and species are more important. For example, the locals used specific parts of specific kinds of plants for particular medicinal purposes. There were not many species used for construction and the trees were used in a species-non-specific manner. In this context, the three settlements are distinctive from each other. Biomass-based uses were far more frequent in Kutajaya than in the other two (Figure 6.3). In terms of total weight harvested, they accounted for 97% in Kutajaya and 80% in Hanjawar, but only 50% in Majasari, where palm sugar accounted for 40%. The people harvested 36 species in total in Hanjawar and 29 in Majasari for biodiversity-use, but only ten in Kutajaya. They used nine species as medicinal plants in Hanjawar, five in Majasari and four in Kutajaya. Low income and remoteness in Hanjawar explain the highest use there. Similarly, people used 21 species for food in Hanjawar and 16 in Majasari, but only 5 in Kutajaya. These findings suggest that the use of NTFPs was more species specific in Hanjawar and Majasari than in Kutajaya.

The locals harvested various items within the national park area, mostly less than 500 m from the park border (Figure 6.4). They did not want to carry the most frequently harvested NTFPs—firewood and fodder plants—very far on account of the weight (Photo 6.3). People in Kutajaya and Hanjawar often visited the wilderness

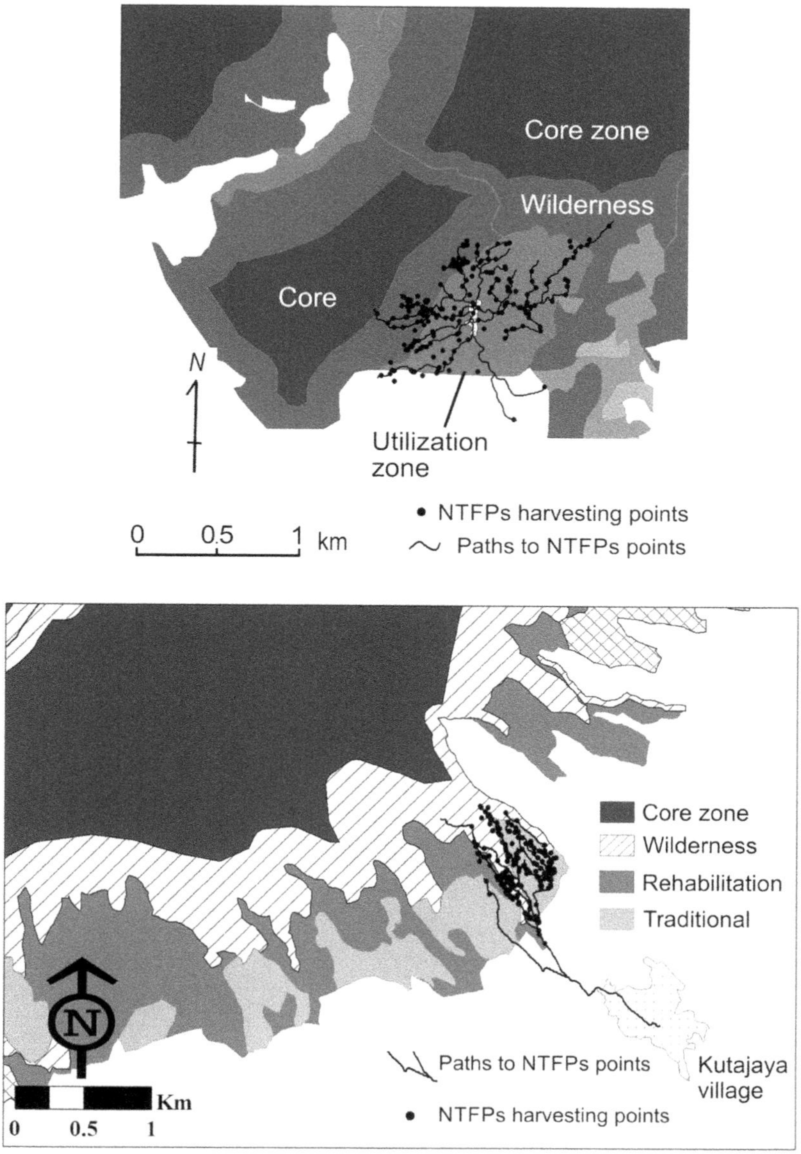

Figure 6.4 NTFPs harvest sites and national park zoning near Kutajaya (*below*), Hanjawar (*above*) and Majasari (*next page*)

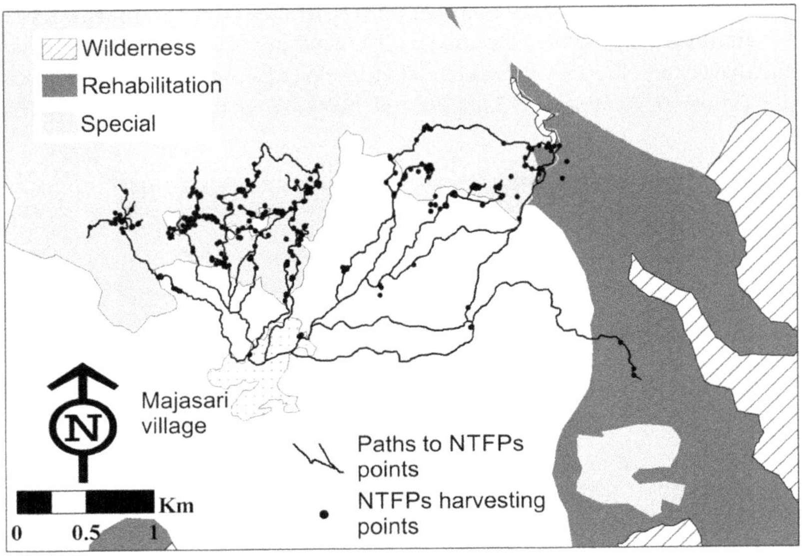

Figure 6.4 (*Continued*)

and rehabilitation zones, where NTFP harvesting was prohibited. According to our interviews, they did not understand the park regulations very well. Relatively well known regulations concerned the bans on cutting trees, hunting wildlife and expansion of farmland. They were dissatisfied with the rules, arguing that they did not build new houses often and so the demand for timber would be too small to disturb forest ecosystems. The park office allowed NTFPs to be harvested for subsistence

Photo 6.3 Firewood and fodder plants collected in the national park.

Photo 6.3 (*Continued*)

living, but did not permit cutting trees. So locals cut down branches, twigs and other small pieces of wood for use as firewood. Park officials did not think that harvesting firewood and fodder plants would have significant impacts on biodiversity. Since villagers collect NTFPs within the park boundaries, communications and discussions between them and national park officers should be essential to resolving conflicts.

6.2.2 *Distribution and abundance of indicator species*

GHSNP selected three animals as indicator species for monitoring biodiversity: Java gibbon *Hylobates moloch*, Javan leopard *Panthera pardus* and Javan hawk-eagle *Spixaetus bartelsi*. Park rangers periodically patrol the park areas to control human activities. We asked them to carry GPS receivers to track the activity of these species, when they sighted the animals or found field signs such as scats, to record the species name and location. From their records we built a database using GIS (Geographic Information System). Then we created an indicator to analyse the distribution and relative abundance of the indicator species according to the observation frequencies per unit of monitoring distance. Leopards were relatively rare, so we set up 31 camera points to record their distribution in the central part of the park, as described below. The observation records indicated that all species were most frequently found near the boundaries of the park. This finding suggests that areas near the boundaries might be frequently used habitat. However, as the rangers patrolled the border areas most frequently, information obtained from areas farther from the boundaries was insufficient for habitat analysis.

Then we sought to analyse the habitat preference of each species in terms of the distance from the park boundaries. Figure 6.5 shows the number of indicator species

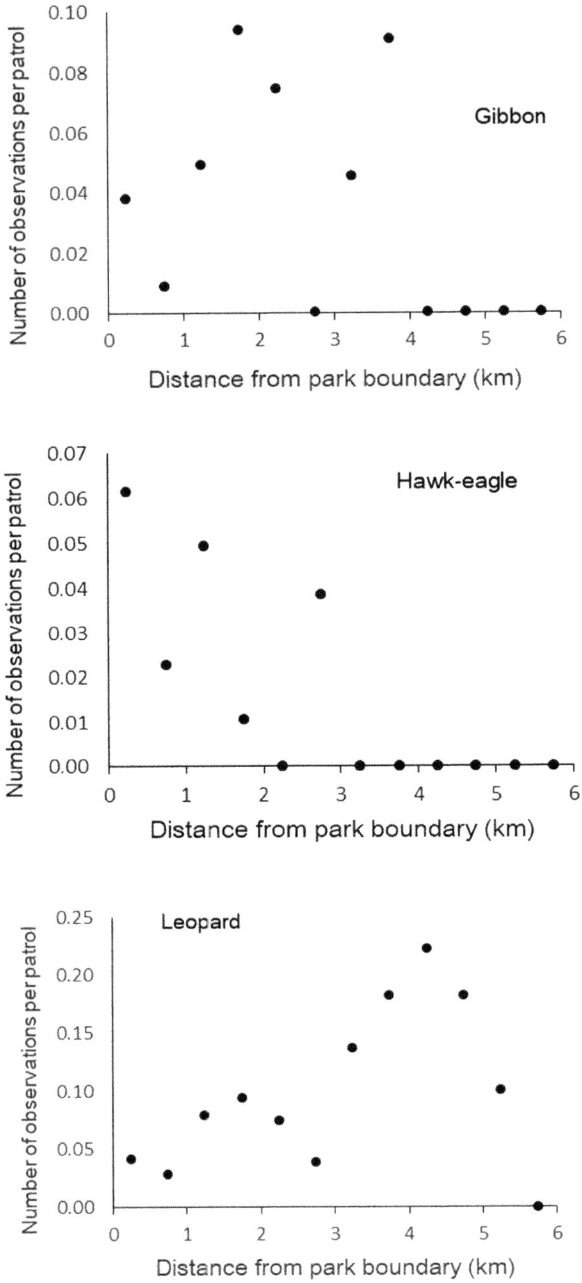

Figure 6.5 Number of indicator species observed per monitoring visit (vertical axis) with respect to the distance of the observation point from the park boundary (horizontal axis); gibbon (*above*), hawk-eagle (*middle*) and leopard (*below*).

observed per unit of monitoring frequency versus distance from the park border. The figures indicate that hawk-eagles appeared more frequently near the border than leopards and that gibbons seemed to prefer intermediate locations. The areas near the park borders were covered mostly by secondary forests, plantations, forest garden and grassland, because the park management has reserved those areas to restore farms to natural vegetation. In contrast, the interior of the park was protected even before the park was established, and remote forests are less frequently used by locals. These results suggest that the habitats of the three species partially coincided with the areas of local people's use, particularly for the hawk-eagles, which favour more disturbed areas.

The Javan leopard is a protected species in Indonesia. Habitat loss and fragmentation, conflicts with locals and poaching are considered to be the main factors threatening its continued existence. We conducted a camera trap survey using 31 cameras (Bushnell Trophy Cam) placed within 655 hectares off forests near Cikaniki village in November 2012. Dones Rinaldi at Bogor Agricultural University designed the survey. Over 720 trap-days, 277 images were collected, featuring 22 species of mammals. Leopards appeared most frequently along with their prey species. The cameras also recorded a wild cat species, ***Prionailurus bengalensis***, whose two subspecies rank highest in the Japanese Red List. A camera placed about 3 km away from Cikaniki captured leopard images most frequently. The cameras that captured leopards were significantly further from the park boundaries than the cameras that did not capture leopards (*t*-test, $P<0.05$). Diameter at breast height of the canopy trees did not significantly correlate with the number of leopards trapped. Therefore, leopards seem to prefer living far from settlements to living in relatively untouched forests.

This study suggests that the areas near the park perimeter, where locals harvested NTFPs for daily living, are also parts of the indicator species' habitats. Figure 6.6 proposes effects of collecting NTFPs and expanding park areas on

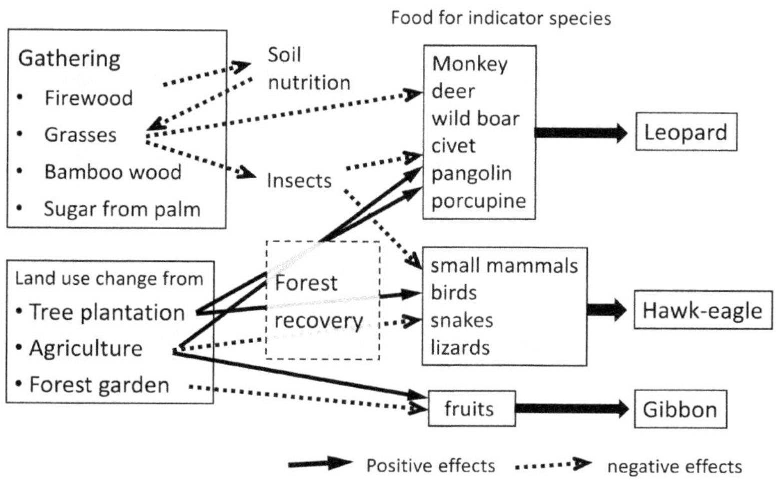

Figure 6.6 Plausible effects of human activities on indicator species.

those habitats. Collecting firewood would not directly affect the food source of the indicator species. Among the prey species of the leopards, only barking deer *Muntiacus muntjak* feed on grass and herbs, which local people gather for livestock. Our camera trapping suggests that human usage is unlikely to cause significant adverse effects on deer owing to their low population density. Besides, collecting forage plants within forest was limited to areas near the boundaries in the dry season, whereas the main leopard habitats were situated further away. Therefore, harvesting NTFPs would not have a significant effect on leopard populations. The hawk-eagle appears to be a typical satoyama species with habitat preferences similar to the grey-faced buzzard *Butastur indicus* in Japan. According to Prawiradilaga *et al.* (2001), it is more dependent on arboreal species for food than the buzzard. Since we often saw hawk-eagles flying over paddy fields, it may be more dependent on reptiles than the above study suggests. Habitat conditions for the gibbon would be improved as long as trees grow larger in the rehabilitation zone and satoyama landscapes are maintained. Gibbons are arboreal and primarily fruit-eaters, and thus forest restoration would give them better habitats.

This study suggests that moving the boundaries between the special zone and the more strict protection zones tailored to the demand for NTFPs will alleviate conflicts between local use and conservation. At the same time, adequate biodiversity conservation should require continued monitoring of the indicator species and other key species as well as limiting human use to areas near the park boundaries.

6.3 Nationwide perspectives

6.3.1 *Biodiversity and its regional evaluation*

Biodiversity is an abstract concept, and its definition varies among biologists (Takacs, 1996). A typical textbook description explains that biodiversity encompasses everything from the level of genetic variance to the diversity of landscapes, as I illustrated in Figure 6.7. It is impractical to deal with such a complex multi-layered reality, from genetic diversity to landscape diversity. Therefore, biologists typically do not use the term biodiversity when publishing papers based on field work.

Species are often considered the fundamental unit of conservation of terrestrial ecosystems and biodiversity. This book focuses on birds and mammals in forest ecosystems, for which information is the best available, comparing species across regions.

It most often refers to bird species diversity, based on the idea that bird species richness would be an appropriate indicator in Japan, where forests make up about two-thirds of the landmass. In addition, bird species diversity would represent various feeding and nesting habitat types, including different elements of vertical forest structure (from the ground to the top canopy layer) and different stages of forest succession (MacArthur and MacArthur, 1961; Nicholas *et al.*,

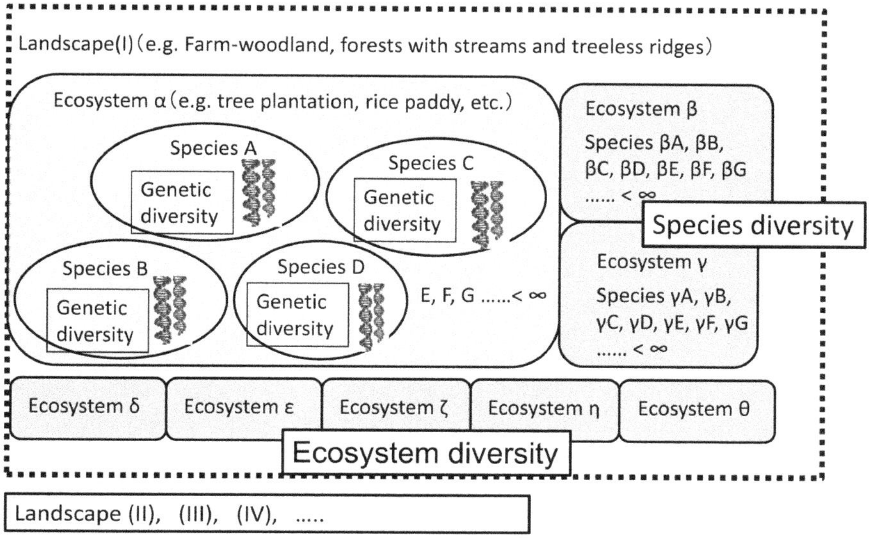

Figure 6.7 Textbook description of "biodiversity".

1999). A high level of bird species diversity would indicate a high diversity of food sources in the ecosystem, such as vertebrates, invertebrates, fruits and nuts. It is crucial to include mammals, because there are far more mammals endemic to Japan than birds, as shown in Chapter 1. It is also our instinct as *Homo sapience*, mammals ourselves. In conservation efforts in general, many people give birds and mammals higher priority than other groups, except for vascular plants. Therefore, the two groups are adequate sources of information as far as forest biodiversity is concerned. Because a particular land management or conservation policy can affect the preservation of other species also, "biodiversity" can encompass many species of conservation concern. For example, when I propose a particular biodiversity conservation policy based on avian data, I expect that it would benefit other critical taxonomic groups as well.

Species diversity can be evaluated from two perspectives; (a) species richness (number of species) and (b) presence of threatened species. The first approach counts the number of species, aiming to protect as many as possible (e.g., Schmidt *et al.*, 2006; Ara'ujo *et al.*, 2007; Pearman and Weber, 2007). One of the problems with this approach is that conservation efforts may be devoted to species that are doing well, in other words, widely distributed, common in disturbed habitat and found in large numbers. It may also be impossible to set aside large enough areas to protect all species. Another problem is that a limited number of richness hotspots, selected on the basis of species richness, may fail to protect many rare species (Prendergast *et al.*, 1993; Curnutt *et al.*, 1994; Dobson *et al.*, 1997; Lawler *et al.*, 2003). The second approach gives arbitrarily weighted scores to threatened species, most commonly those on the Red List (e.g., Stattersfield, 1998; Pearman, 2002).

This approach gives a score to a species according to its status on the list. Then it sums the scores of each species to obtain an evaluation value for a given area to compare with other areas. The evaluation is dependent on the number of such species and the status of each species on the list. This measurement is simple and clear. Nevertheless, it may exclude areas that are rich in species but do not have highly endangered species, since species-rich hotspots and endangered-species hotspots do not always coincide (Neel and Cummings, 2003; Drummond *et al.*, 2010).

To examine the difference between the two approaches and regional variation in Japan, I used the same database as the one introduced before (refer to Supplementary Note 6-3a). Among the species included in a total of 2899 grid cells, this analysis focused on 121 species for which forests are an essential habitat component (Tojo, 2007). The first approach evaluated species richness in terms of the number of species observed in each cell. The second approach evaluated each cell on the basis of threatened species, arbitrarily weighted according to the national Red List status: critically endangered (CR, weighted 4), endangered (EN, 3), vulnerable (VU, 2) and near threatened (NT, 1). Then it compared the geographical distribution of hotspots selected according to these two methods.

The correlation between species richness and the scores based on the Red List status was low (Figure 6.8). Many grid cells were abundant in threatened species but low in species numbers and vice versa. There was no coincidence between the ten most species-rich cells and the ten cells with the highest threatened species scores. Among the cells with the highest scores, the distribution of 77 species-rich hotspots and 83 threatened-species hotspots differed (Table 6.8). The former were abundant in central and north-eastern Japan, and the four uppermost regions in the table accounted for 83% of species-rich hotspots. In

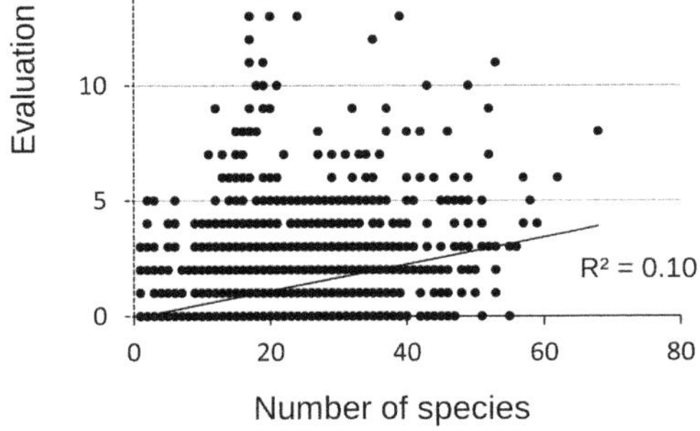

Figure 6.8 Correlation between species richness and scores based on the national Red List categories.

Table 6.8 Regional comparison of the number of species-rich grid cells and cells with high threatened species scores

Region	*Species-rich*	*Threatened species*
Hokkaido	14	1
Tohoku (northern Honshu)	22	6
Kanto (mid-eastern Honshu)	8	4
Chubu (central Honshu)	20	13
Kinki (mid-western Honshu)	3	0
Chugoku (westernmost Honshu)	5	5
Shikoku	1	3
Kyushu	4	12
Nansei Archipelago	0	32
Izu and Ogasawara Islands	0	7
Total	77	83

The Izu Islands are small islands that run nearly in a north-south direction from the south of the Kanto region to the north of the Ogasawara Islands.

contrast, Kyushu and remote islands in the south had many threatened-species hotspots, accounting for 61% of the total. Only the Chubu region had many species-rich and threatened-species hotspots. There, southern and northern species may coexist along a wide altitudinal range from sea level to as high as 3000 m, the highest mountains in the country. As indicated in Chapter 1, northern Japan has higher proportions of natural forests than southwestern Japan apart from the Nansei Archipelago (Table 1.3). Such a difference may cause regional variation in species richness, as fewer species were found in plantations in northwestern Kyushu than in the other case study areas. The uniqueness of the remote islands is described in Chapter 2 and is also shown in Table 6.8.

In contrast to northwestern Kyushu, there were many grid cells with high scores for threatened species in Kyushu (ten species in total). Eleven of the twelve cells were in central and southern Kyushu, where rare species of spring migrants were found. Five species of these migrants do not fly north and have small populations, so they are ranked high on the national Red List, but none of them is endemic to Japan. Few studies have compared satoyama and okuyama in Kyushu, but according to various unpublished documents and my field experience, the main habitat for at least eight of the ten threatened species is satoyama. Satoyama in Kyushu may be richer in species diversity than okuyama, where plantations are predominant except in high-altitude mountainous areas.

The results generally correspond to the results obtained in the four regional case studies. Okuyama in Tadami is a species-rich place often found in northern Japan. The species richness in the satochi in Tsukuba relies more on the combination of a variety of landscape elements than satoyama forest management. The relatively species-poor habitat in northwest Kyushu may have conservation value for spring migratory birds. Combined with the results on the mammals, birds and other terrestrial vertebrates in Chapters 2–5, the need for the following strategies at the national level are indicated. Highest priority should be given to the endemic species on remote islands. The second priority is okuyama. These two together cover the majority of endemic and endangered wildlife species in the country. The third priority is satoyama in areas where okuyama is dominated by plantations. The coexistence of rice paddies and secondary forests is critical in satochi and probably in satoyama landscapes. However, no conservation movements are evident in northwestern Kyushu at the moment.

6.3.2 *Roles of protected areas and biodiversity-based ecosystem services*

Species diversity is widely used as a conservation target, and protected areas are recognised as effective in preserving it (Mathur *et al.*, 2011). However, biodiversity conservation remains a relatively low priority in land use planning. In Japan, strictly protected areas (the special protection and first-class special zones of national parks) cover only 1.3% of the national land area, according to the statistics provided in Chapter 1. In contrast, plantations, including those created by clearing many of the last remaining old-growth forests on marginal lands, now cover 28% (41% of the forested area) (Table 1.3). Adding or expanding protected areas to improve conservation is difficult, primarily because the Ministry of the Environment owns just 0.3% of the total land area of Japan's national parks. The term biodiversity appeared for the first time in the Natural Park Act in 2002, associated with a clear mission to protect and restore biodiversity. Even after this change, forest biodiversity needs to be maintained in the natural forests outside the strictly protected areas, as explained in the preceding chapters.

In Indonesia, in contrast, land ownership is not a critical issue, as the land belongs to the competent ministry, which realises the importance of biodiversity conservation. There are some illegal activities, but their impacts on wildlife conservation are minor in GHSNP. In Japan large areas of parkland belong to the Forestry Agency, which does not recognise biodiversity conservation as one of its primary missions. The special protection and first-class special zones usually occur in the high mountains, where timber production is considered economically unfeasible. The Forestry Agency conducts forestry in the other zones. Since large areas are suitable for growing sugi on the mountains of Kyushu, there are only small areas of strict protection in the national parks. Thus, generally speaking, habitat suitability for commercial timber trees has determined the balance between conservation and forestry in Japan, as also explained throughout the preceding chapters.

In addition to land tenure, another factor in the role of protected areas is the number of rangers in national parks. GHSNP has over a hundred rangers who not only monitor species and control illegal activities, but also negotiate with locals. Japanese national parks have only a few rangers, of whom one or two are fixed-term employees. Another, albeit relatively minor, factor is the lack of well-educated and specialised personnel trained within the fields of ecology and conservation in Japan. Relatively few universities have teach these fields in Japan, and some of the top-ranking universities are in the Kinki region (Kyoto, Osaka and Kobe) where there were no hotspots with high threatened species scores in the bird study (Kinki does not appear in Table 6.8). The Ministry of the Environment did not employ any biologists for many years after its establishment in favour of landscape architects. This lack of disciplines may be one of the reasons why the ministry has kept arguing that satoyama should play the principal role in biodiversity conservation, almost overlooking the importance of okuyama landscapes.

National parks and forest reserves played a small role in biodiversity conservation in the case study areas. Amami suffered from extensive clear-felling of old forests and mongoose invasion. The government established the national park only after they were no longer major issues. Before its establishment, there was a quasi-national park that had only a small strictly protected area of about 100 hectares around the summit of the island's highest peak, which did not provide suitable habitats for many endangered species. Amami was fortunate that logging became economically unfeasible in the early 1990s owing to the globalisation of the forestry economy and Japan's strong economic power at that time; otherwise, loggers would have cleared almost all the old forests, leaving only small fragments. In Tadami, forest conservation had a different background from that on Amami: The national forest areas are much larger than on Amami; the human population is less than one-tenth that on Amami; and the leadership of a few individuals has set nature conservation as one of the town's primary policy goals. Unlike on Amami, the flora and fauna in Tadami did not capture the attention of the whole country. The regional office of the Forestry Agency gave up logging all the natural forests after a local conservation group revealed illegal logging. These effective conservation initiatives were primarily local. The other effective factor was deep snow that prevented most of the natural forests from being converted into plantations. Still, Amami and Tadami have two similarities: The Forestry Agency established protected areas after cutting most of the natural forests for chip wood; and the management plans of the broadleaf forests were changed according to the global economy and the fiscal deficit of the agency.

There are no protected areas in the satochi-satoyama landscapes of Tsukuba. The small reserve around the top of Mt Tsukuba resembles okuyama with large trees. At the same time, a wide range of habitat types and altitudes has maintained high levels of species richness. Strong conservation measures may not be necessary, as the plants and animals have already survived high levels of habitat disturbance. What is currently needed is to prevent further fragmentation of the natural forests in satochi. In contrast to these three regions above, Nagasaki Prefecture has two old national parks that are not very concerned with biodiversity conservation. These parks were established primarily for nature tourism, and, according to the databases and literature, have no species of mammals or birds on the national Red List.

There are two types of endangered species in terms of the extent of human impacts in wooded landscapes. One lives in landscapes that are disturbed naturally but that are not disturbed much by human activities. The case studies on Amami and in Tadami, in which protected areas may play a critical role in preserving old forests with large-diameter trees, introduced this type of endangered species. There are far more endangered and endemic species of this type than in the other study areas, at least among terrestrial vertebrates. The other type of endangered species can live in relatively disturbed satoyama landscapes. Thus, it would be necessary to protect already fragmented natural forests from land use changes.

Biodiversity provides various ecosystem services with a wide range of regional variations. Among the regions compared, forests in Tadami provide the broadest range of services; harvesting NTFPs, angling and nature tourism. Local residents enjoy recreational activities frequently in satoyama landscapes that have high levels of bird species diversity. As shown in Table 5.1, the main purpose of satoyama activities relates biodiversity-based services, such as nature observation, biological survey and environmental education. The coexistence of biodiversity and ecosystem services would play an essential role in satoyama landscapes. In Japan, there is no national policy strategy to secure and manage the landscapes as essential areas for biodiversity-based services. In addition to scenic beauty and biodiversity conservation, it is time to consider such services for the sake not only of plants and animals but, also of people.

6.4 Summary discussions: wildlife, landscape services and society

Certainly, Japan is very rich in biodiversity, and people enjoy the natural landscape. However, it is not the product of successful nationwide conservation efforts. On the other hand, the various circumstances surrounding biodiversity conservation have changed over time, as illustrated so far. In particular, regional differences in such changes are remarkable.

Until the early 1990s, forest managers paid little attention to the species unique to Amami and believed that logging would have negligible effects on their populations. Only a few scientists visited the island for research. Around 2000 many tourists visited the pristine forest not far from the urban area of Amami City. They visited the island to enjoy diving but showed an interest in rare flora and fauna. The late 1990s was the turning point after the Rio de Janeiro global environmental summit, when the Japanese government enacted several laws on biodiversity conservation. The Forestry Agency was forced to acknowledge that environmental conservation should be as essential as timber production, mainly because of its enormous budget deficit, even though a minority still think that clear-cutting would have insignificant effects on rare species. Amami is an exceptional case in that not only biologists but also many tourists from across the county realise the importance of the unique species.

Tadami Town is an example of a local government that has expressed its enthusiasm for nature conservation and decided to revitalise the local economy with extensive beech forest ecosystems and NTFPs. The turning point came in the early 2000s when a few people sued against illegally logged national forests. At that time, the local office of the national forest already recognised that the policy should be changed from the idea that timber should take precedence over nature conservation. Thus, the movements in Tadami were more local than those on Amami. Collection of NTFPs in the mountainous areas is also unique to Tadami, and is deeply rooted in the daily lives of the local people, who continued to go out even after the power plant accident.

Satoyama landscapes have retained rich species diversity even after the decline of traditional landscape management since around 1960 during the period of rapid economic growth. It is likely that species diversity has decreased, although no landscape-scale studies have revealed any significant change. However, the national and local governments have not taken any specific conservation measures to restore coppice management and hay fields, as well as the traditional custom of flooding rice paddies all year round. At the national level, no conspicuous extinction events occurred in satoyama in recent years. The Japanese crested ibis became extinct in the 1970s in satoyama, but the primary cause was killing as pests and not the loss of traditional landscapes (Chapter 1). On the other hand, local governments support satoyama activities and seem to set biodiversity as one of their activity targets. Forests under such activities are too small to have significant effects on regional biodiversity. The most significant role of satoyama biodiversity in society would be the provision of recreational opportunities. Satochi landscapes and relatively tall mountains near urban areas described in the last chapter are typical examples. The human-nature interactions in satoyama are significantly different from those on Amami and those in Tadami, and it seems that past landscape changes in satoyama have not adversely affected modern nature-based recreation.

In northwestern Kyushu, landscape changes have occurred in okuyama, such as the expansion of tree plantations. Although there are several endangered species of terrestrial vertebrates in satoyama in northwestern Kyushu, biodiversity is not as high as in the other study areas. Forest managers and the general public there are not very interested in nature conservation and nature tourism is less popular there than in the other study areas. The primary role of forests has been timber production. In the 1990s and 2000s, increasing interest in biodiversity conservation at the national level and more prominent issues led to fundamental changes in forest policy. However, the above circumstances in northwestern Kyushu have not changed.

The above comparison suggests that natural landscapes are frequently used in various ways in areas of relatively high wildlife species diversity. The extent of social interests in biodiversity seems to correlate with the uniqueness of the fauna. Interactions between species richness and society seem to affect the regional characteristics of the behaviours of local people and the attitudes of local governments towards natural landscapes and biodiversity issues. When local society recognise high conservation values, governments eventually change their strategies, significantly reducing the conflicts between forestry and conservation.

In contrast to biodiversity-based ecosystem services, economics of local and global scale had profound impacts on forest management policies and biodiversity conservation in the past. The rise of the Japanese yen against foreign currencies stopped clear-cutting of old forests on Amami. Loggers felled as many natural forest stands as possible and forest owners planted as many commercial trees as economically viable with government subsidies. However, the situation turned for better after the Rio Summit in 1992, and the government has enacted several laws on biodiversity. The environmental impact assessment process is supposed to take into account the national and local Red List species. Therefore, economic efficiency may have to compromise with biodiversity. An even more significant factor is the massive population declines that are projected to occur in rural areas within the coming few decades. Projected decline in Tadami (Chapter 4) will occur similarly throughout the country. While many rural societies have regarded forestry vital for employment, population declines would result in reducing silvicultural management and facilitate replacement of plantations with naturally regenerated forests. In addition, the government budget for civil engineering works is decreasing due to the rapidly ageing population and increasing need for social security, which will hinder the construction and maintenance of forestry roads.

The most rapidly ageing society in the world will reduce human impacts in the future and benefit biodiversity conservation. Meanwhile, it is projected that the metropolitan areas will grow in population over the next one to two decades, increasing the demand for nature tourism and further enhancing the satoyama activities. Such changes are beneficial for increasing concern for biodiversity and interest in biodiversity-based services. In addition, the enormous government budget deficit will reduce the number of devastating civil engineering projects in the rapidly depopulating mountainous regions in the future.

Nevertheless, there are many fundamental social issues to be resolved in order to have better conditions for biodiversity conservation and biodiversity-based services in each region. Some improvements at the national level are apparent, such as improving conservation education, transfer of jurisdiction over the national land in the national parks to the Ministry of the Environment and an increase in the ratio of conservation budget to development budget. From an academic perspective, it is crucial to increase the number of qualified studies that integrate multifaceted socio-ecological aspects with system approaches.

Supplementary notes

6-1) *Shinrin ringyō hakusho (Heisei 30-nendo).* Available at: ñhttps://www.rinya.maff.go.jp/j/kikaku/hakusyo/index.html (accessed 16.07.19).

6-2) The information on species distribution is based on *Dōbutsu bunrui-gun-betsu bunpuzu.* Available at: https://www.biodic.go.jp/kiso/atlas/pdf/3.mammals.pdf (accessed 15.04.20).

6-3) a) *Dai-6-kai shizen kankyō hozen kiso chōsa chōrui hanshoku bunpu chōsa hōkokusho.* Available at: http://www.biodic.go.jp/reports2/6th/6_bird/

index.html (accessed 22.04.20). This database gives the distribution of breeding birds on grids of approximately 10 km^2, based on systematically designed transect lines and questionnaire surveys conducted by volunteers (Nihon yachō no kai) from 1997 to 2002. The volunteers recorded all species from 1 April to 31 August, covering the entire breeding season from northern to southern Japan. Most raptors are excluded to protect their breeding sites from bird watchers. Amano and Yamaura (2007) described the survey in more detail.

b) *Monitaringu saito 1000* (Kankyō-shō seibutsu tayōsei sentā) Available at: http://www.biodic.go.jp/moni1000/findings/data/index.html (accessed 16.04.20). The report is based on a 500 m route survey, excluding the raptors, along 17 routes scattered in northwestern Kyushu.

6-4) Available at: https://www.env.go.jp/nature/satoyama/field.pdf (accessed 25.12.18).

6-5) Available at: https://www.e-stat.go.jp/stat-search/files?page=1&bunya_l=02 (accessed 25.12.18).

6-6) Available at: https://www.jalan.net/kankou/ (accessed 25 and 26.04.20).

The rankings are based on the number of votes and evaluation by the visitors to each site. The determination process is not stated in detail.

6-7) *Nōsakubutsu higai-jōkyō*. Available at: https://www.maff.go.jp/j/seisan/tyozyu/higai/hogai_zyoukyou/index.html (accessed 28.04.20).

References

Abe, H. (ed.) (2008). *Nihon no honyūrui*. Hadano: Tokai-daigaku Shuppankai.

Abe, S., Handa, Y. and Takatsuki, Y. (1991). 1990-nen 1-gatsu kara 1991-nen 1-gatsu madeno Amami Ōshima niokeru mangūsu no hokaku kekka. *Chirimos*, 2, pp. 27–57.

Akimichi, T. and Morimoto, Y. (2012). 'Western Japan cluster', in Duraiappah, A.K., Nakamura, K., Takeuchi, K., Watanabe, M. and Nishi, M. (eds.), *Satoyama-satoumi ecosystems and human well-being: socio-ecological production landscapes of Japan*. Tōkyō: United Nations University Press, pp. 354–380.

Amano, T. and Yamaura, Y. (2007). Ecological and life-history traits related to range contractions among breeding birds in Japan. *Biological Conservation*, 137 (2), pp. 271–282.

Anon. (1977). *Tokubetsu ten'nen kinenbutsu Amami-no-kurousagi no genjō: ten'nen kinenbutsu kinkyū hōkoku*. Kagoshima: Kagoshima-ken Kyōiku Iinkai.

Aoyagi, M. (1991). Shutoken kinkō no higaeri recreation-chi no juyō hyōka nikansuru kenkyū. *Nōson Keikaku Gakkai-shi*, 10 (1), pp. 7–16.

Ara'ujo, M., Lobo, J.M. and Moreno, J.C. (2007). The effectiveness of Iberian protected areas in conserving terrestrial biodiversity. *Conservation Biology*, 21 (6), pp. 1423–1432.

Atkinson, G., Bateman, I. and Mourato, S. (2012). Recent advances in the valuation of ecosystem services and biodiversity. *Oxford Review of Economic Policy*, 28 (1), pp. 22–47.

Balvanera, P., Pfisterer, A.B., Buchmann, N., He, J.S., Nakashizuka, T., Raffaelli D. and Schmid, B. (2006). Quantifying the evidence for biodiversity effects on ecosystem functioning and services. *Ecology Letters*, 9 (10), pp. 1146–1156.

Bond, R.R. (1957). Ecological distribution of breeding birds in the upland forests of southern Wisconsin. *Ecological Monographs*, 27, pp. 351–384.

Chapman, J.A. and Flux, J.E.C. (1990) 'Introduction and overview of the lagomorphs', in Chapman, J.A. and Flux, J.E.C. (eds.) *Rabbits, hares and pikas: status survey and conservation action plan*. Gland: IUCN, pp. 1–6.

Conservation International. (2011). *Biodiversity hotspots revisited*. Conservation Synthesis, Center for Applied Biodiversity Science at Conservation International. Available at: http://www.biodiversityhotspots.org/xp/Hotspots/resources/maps.xml (accessed 9.07.18).

Curnutt, J., Lockwood, J., Lu, H.-K., Nott, P. and Russell, G. (1994). Hotspots and species diversity. *Nature*, 367, pp. 326–327.

Diamond, J.M. (1975). The island dilemma: lessons of modern biogeographic studies for the design of natural reserves. *Biological Conservation*, 7 (2), pp. 129–146.

Dai-jū kanku kaijō hoan honbu kaiyō jōhō. (2018). *Ocean surface temperature Kaishō jōhō*. Available at: https://www1.kaiho.mlit.go.jp/KAN10/kaisyo.html (accessed 25.09.18).

Dobson, A.P., Rodriguez, J.P., Roberts, W.M. and Wilcove, D.S. (1997). Geographic distribution of endangered species in the United States. *Science*, 275, pp. 550–553.

Duraiappah, A.K., Nakamura K., Takeuchi K., Watanabe M. and Nishi M. (eds.) (2012). *Satoyama-satoumi ecosystems and human well-being: socio-ecological production landscapes of Japan*. Tōkyō: United Nations University Press.

Drummond, S.P., Wilson, K.A., Meijaard, E., Watts, M., Dennis, R., Christy, L. and Possingham, H.P. (2010). Influence of a threatened-species focus on conservation planning. *Conservation Biology*, 24 (2), pp. 441–449.

Ebihara, J. (2011). 'Nihon no koyū shokubutsu no hotto spotto', in Katō, M. and Ebihara, J. (eds.), *Nihon no koyū shokubutsu*. Hadano: Tōkai-daigaku shuppankai, pp. 17–59.

Elmqvist, T. and Maltby, E. (2010). 'Biodiversity, ecosystems and ecosystem services', in Kumar, P. (ed.), *The economics of ecosystems and biodiversity: ecological and economic foundations*. London: Wageningen University, pp. 41–112. Available at: https://www.researchgate.net/publication/48192189_Chapter_2_Biodiversity_ecosystems_and_ecosystem_services (accessed 10.07.19).

Fa, J.E. and Bell, D. (1990) 'The volcano rabbit', in Chapman, J.A. and Flux, J.E.C. (eds.) Rabbits, hares and pikas: status survey and conservation action plan. Gland: Switzerland, pp. 141–143.

Fa, J.E., Romero, F.J. and Lopez-Paniagua, D.J. (1992). Habitat use by parapatric rabbits in a Mexican high-altitude grassland system. *Journal of Applied Ecology*, 29, pp. 357–370.

Forman, R.T.T., and Godron, M. (1986). *Landscape ecology*. New York: John Wiley & Sons.

Fujita, Y. (1984). *Gendai nihon no shinrin mokuzai shigen mondai*. Tōkyō: Sekibunsha.

Fukasawa, K., Hashimoto, T., Tatara, M. and Abe, S. (2013). Reconstruction and prediction of invasive mongoose population dynamics from history of introduction and management: a Bayesian state-space modelling approach. *Journal of Applied Ecology*, 50 (2), pp. 469–478.

Fukui, A., Yasuda, M., Kamiyama, K. and Kanai, Y. (2005). Zenkokutekina chōrui chōsa "tori no seisoku kankyō monitaringu chōsa" de akirakani natta hanshokuki no chōrui gunshū no shukōsei. *Strix*, 23, pp. 1–29.

Fukushima-ken Nōrinsuisan-bu. (2018). *Heisei 30-nen Fukushima-ken shinrinringyō tōkeisho*. Available at: https://www.pref.fukushima.lg.jp/sec/36055a/toukeisyo.html (accessed 15.12.19).

Haines-Young, R. and Potschin, M. (2010). 'The links between biodiversity, ecosystem services and human well-being', in Raffaelli, D. and Frid, C. (eds.) *Ecosystem ecology: a new synthesis*. Cambridge: Cambridge University Press, pp. 110–139.

Harris, L.D. (1984). *The fragmented forest: island biogeography theory and the preservation of biotic diversity*. Chicago and London: University of Chicago Press.

Harrison, P.A., Berry, P.M., Simpson, G., Haslett, J.R., Blicharska, M., Bucurd, M., Dunford, R., Egohe, B., Garcia-Llorente, M., Geamănă, N., Geertsema, W., Lommelen, E., Meiresonne, L. and Turkelboomi, F. (2014). Linkages between biodiversity attributes and ecosystem services: a systematic review. *Ecosystem Services*, 9, pp. 191–203.

Hasumi, O. (1981). 'Amami niokeru nōgyō no henka to nōson', in Matsubara, J., Toya, O. and Hasumi, O. (eds.), *Amami nōson no kōzō to hendō*. Tōkyō: Ochanomizu Shobō, pp. 19–55.

Hayase, N. and Yamane, S. (1982). Ibaraki-ken Tsukuba sankei niokeru hakone sanshōuo *onychodactylus japonicas* no suichū seikatsuki no seitai. *Nihon Seitai Gakkaishi*, 32 (3), pp. 395–403.

Hayashi, Y. and Suzuki, H. and Kirino, M. (1984). 'Fun kara suisoku sareru kurousagi no kōdō', in WWF Japan Kagaku-iinkai (ed.) *Nansei Shotō to sono shizen hogo*. Tōkyō: WWF Japan, pp. 293–302.

Hiura, T., Fujiwara, K., Hōjyō, H., Okada, J., Udō, H., Okuyama, S., Morita, H., Fukuta, H., Fujito, E., Fukui, T., Takahata, M., Arikura, K., Sugiyama, H. and Takeda, T.

(1995). Hokkaidō-daigaku nakagawa-chihō enshūrin niokeru gensei hozonrin no shinrin kouzō to sono chōki dōtai. *Hokkaido Daigaku Enshūrin Hōkoku*, 52 (2), pp. 85–94.

Hoth, J., Valazquez, A., Romero, F.J., Leon, L., Aranda, M. and Bell, D.J. (1987). The volcano rabbit—a shrinking distribution and a threatened habitat. *Oryx*, 21 (2), pp. 85–91.

Nihon Hachū-ryōseirui Gakkai. (2019). *Nihonsan hachū-ryōseirui hyōjun wamei listo, 2019-nen 6-gatsu 26-nichi ban*. Available at: http://herpetology.jp/wamei/pdf_ja.php (accessed 6.07.19).

IAEA (2006) 'Environmental consequences of the Chernobyl accident and their remediation: twenty years of experience', in *Report of the Chernobyl forum expert group. Environment, radiological assessment reports series*. Vienna: International Atomic Energy Agency.

Ibaraki-ken Shizen-hakubutsukan. (1998). *Dai-ichiji sōgō chōsa hōkokusho*. Available at https://www.nat.museum.ibk.ed.jp/publications/3/publications_17.html (accessed 1.01.20).

Ikeda, S., Kaneko, S., Akama, A. and Takahashi, M. (2014). Shinrin dojō no hōshasei seshiumu bunpu to dōtai no chōsahō. *Shinrin Sōgō-kenkyūsho Kenkyū Hōkoku*, 432 (3), pp. 137–145.

Ikehara, S. (1980). 'Tōyō no Galapagosu', in Kizaki, K. (ed.), *Ryūkyū no shizenshi*. Tōkyō: Tsukiji Shokan, pp. 86–100.

Ikehara, S. (1989) Kaki to tōki no Amami Ōshima niokeru Amaminokurousagi no mokugeki kiroku. in Anon. (ed.), *Shōwa 63 nendo Amami Ōshima chōsa hōkokusho*. Tōkyō: Kankyō-chō Shizen Hogo-kyoku, pp. 131–143.

Ikehara, S., Abe, T., Chinen, M., Yonashiro, Y., Chigira, Y., Higoshi, K. and Mitsui, K. (1981). *Kenaganezumi jittai chōsa hōkokusho*. Naha: Okinawa-ken Kyōiku-iinkai.

Imaizumi, Y. (1960). *Genshoku nihon honyūrui zukan*. Ōsaka: Hoikusha.

Imazeki, R., Ōtani, Y. and Hongō, T. (2011). *Nihon no kinoko*. Tōkyō: Yama-to-keikoku-sha.

Inagaki, H., Saiki, C., Matsuno, K. and Ichihara, M. (2015). Suiden ni okeru akasujikasumikame no kotaigun hendō ni kankeisuru seitaikei sābisu no rei. *Nihon Seitai Gakkaishi*, 65 (3), pp. 291–298.

Inoue, T. (2019). *Nihon rettō kaihukuron*. Tōkyō: Shinchō Senshe.

Ishii, N. (2003). Amami Ōshima no mangūsu kujo jigyō: tokuni seisokusū no suitei to kujo kōka nitsuite. *Hozenseitaigaku Kenkyū*, 8 (1), pp. 73–82. DOI: 10.18960/hozen.8.1_73.

Ito, T. (1996). Influence of forestry on the formation of national park policy in Japan. *Journal of Forest Planning*, 2 (2), pp. 85–95. Available at: https://doi.org/10.20659/jfp.2.2_85 (accessed 2.09.18).

IUCN. (2019). *Red List; Rock Ptarmigan Lagopus muta*. Available at: https://www.iucnredlist.org/species/22679464/113623562 (accessed 4.07.19).

Iwata, R. (2015). Ryokōsha wa doko kara kurunoka (2) Tsukuba-san no jirei kenkyū. *Tsukubagakuin-daigaku Kiyō*, 10, pp. 23–29.

Jepson, P. and Whittaker, R.J. (2002). Histories of protected areas: internationalization of conservationist values and their adoption in the Netherlands Indies (Indonesia). *Environment and History*, 8 (2), pp. 129–172.

Kaneko, S., Takahashi, M., Ikeda, S. and Akama, A. (2014). Fukushima Genpatsu Jiko niyoru shinrin seitaikei niokeru hōshasei seshiumu osen to sono dōtai. *Nihon Dojō Hiryōgaku Zasshi*, 85 (2), pp. 86–89.

Kankyō-chō. (1999). *Heisei 20-nendo Amami Ōshima niokeru Jawa mangūsu bōjo jigyō hōkokusho*. Naha: Kankyō-shō Naha Sizen Kankyō-jimusho.

Kankyō-shō. (2001). *Kokuritsu kokutei-kōen to hoanrin kokuyurin no chōhukujōkyō*. Available at: http://www.biodic.go.jp/biodiversity/about/initiatives2/files/meeting2/ei2-4-3.pdf (accessed 29.07.19).

Kankyō-shō. (2004). *Honyūrui bunpu chōsa hokokusho*. Tōkyō: Kankyō-shō shizen hogokyoku seibutsutayōsei sentā. Available at: https://www.biodic.go.jp/kiso/15/do_kiso4.html (accessed 19.04.20).

Kankyō-shō. (2010). *Biodiversity of Japan: a harmonious coexistence between nature and humankind*. Tōkyō: Heibonsha.

Kankyō-shō. (2012). *Seibutsutayōsei oyobi seitaikei sābisu no sōgō hyōka hōkokusho*. Available at: http://www.env.go.jp/nature/biodic/jbo2.html (accessed 8.08.18).

Kankyō-shō. (2014). *Heisei 25-nendo Amami Ōshima niokeru Jawa mangūsu bōjo jigyō hōkokusho*. Naha: Kankyō-shō Naha Sizen Kankyō-jimusho.

Kankyō-shō. (2018). *Red List 2018*. Available at: https://ikilog.biodic.go.jp/Rdb/booklist (accessed 4.07.2019 and 15.07.2019).

Kankyō-shō. (2019). *Shizen-kōen menseki no sōkatsu*. Available at: https://www.env.go.jp/park/doc/data/natural/naturalpark_1.pdf (accessed 30.07.19).

Katoh, K., Sakai, S. and Takahashi, T. (2009). Factors maintaining species diversity in *satoyama*, a traditional agricultural landscape of Japan. *Biological Conservation*, 142 (9), pp. 1930–1936.

Kawamitsu, Y., Kawamoto, T., Yoshihara, T. and Murayama, S. (1997). Shokubutsutai-chū no kakushu ion dōtai karamita mangurōbu 3-shu no taiensei no hikaku. *Ryūkyūdaigaku Nōgakubu Gakujutsu-hōkoku*, 44, pp. 91–105.

Kawarasaki, S. and Sugimura, K. (2012). Intārnetto kensaku niyoru sansai to yasei shokuyō kinoko no saishū hindo no suitei to sono chiikisei. *Nihon Shinrin-gakkaishi*, 94 (1), pp. 95–99.

Kimura, K. (2016) Higashi-nihon-daishinsai ga Fukushima-ken Soso-chiho no ringyo mokuzai sangyo ni ataeta eikyo to shinrin ringyo gyosei no torikumi. Ringyō Keizai Kenkyū, 62(3), pp. 68–77.

Kirino, M., Hayashi, Y. and Suzuki, H. (1984). 'Amami-no-kurousagi no shokusei to kōdō patān', in WWF Japan Kagaku-iinkai (ed.) *Nansei Shoto to sono shizen hogo*. Tōkyō: WWF Japan, pp. 303–307.

Kishō-chō. (2019). *Kakushu dēta shiryō*. Available at: http://www.jma.go.jp/jma/menu/menureport.html (accessed 4.07.19).

Kitamura, M. (1998). *Buna no mori to ikiru*. Kyōto: PHP Kenkyu-sho.

Kitamura, S., Ikuta, K., Shikama, T. and Nakamura, H. (2004). Anketo chousa kara mita yukawa niokeru yugyo no jittai. *Suiken Sentā Kenpō*, 12, pp. 1–9. Available at http://tuna.fra.affrc.go.jp/bulletin/bull/bull12/kitamura.pdf (accessed 2.09.18).

Kiyono, Y. and Akama, A. (2013). Radioactive cesium contamination of edible wild plants after the accident at the Fukushima Daiichi Nuclear Power Plant. *Japanese Journal of Forest Environment*, 55, pp. 113–118.

Kizaki, K. and Ohshiro, I. (1980). 'Ryūkyū rettō no oitachi', in Kizaki, K. (ed.), *Ryūkyū no shizenshi*. Tōkyō: Tsukiji Shokan, pp. 8–31.

Kokuritsu Shizen Kagaku Hakubutsukan. (2018). *Nihon no seibutsutayōsei hotto supotto no kōzō ni kansuru kenkyū*. Available at: https://www.kahaku.go.jp/research/activities/project/hotspot_japan/endemic_list/ (accessed 19.01.18).

Koyama, S. and Sugito, S. (1984) Jōmon jinkō simulōshon. Kokuritsu Minzokugaku *Hakubutsukan Kenkyū Hōkoku*, 9(1), pp. 1–39.

Krebs, C.J., Gilbert, B.S., Boutin, S. and Boonstra, R. (1986) Estimation of snowshoe hare population density from turd transects. Canadian Journal of Zoology, 65(3), pp. 565–567.

Kubo, H. and Supriyanto, B. (2010). From fence-and-fine to participatory conservation: mechanisms of transformation in conservation governance at the Gunung

Halimun-Salak National Park, Indonesia. *Biodiversity and Conservation*, 19 (6), pp. 1785–1803.

Kuboyama, H. (2018). Mokushitsu baiomasu hatsuden no genjō to kadai. *Kankyō Jōhō Kagaku*, 47 (2), pp. 10–15.

Kunimitsu, Y. (2005). Jūmin no riyō jōkyō kara mita shinsui kōen no yūchi ken'iki nikansuru kenkyū: kakuritsu-gata gravity model no shisetsu keikaku eno tekiyōkanōsei. *Nōson Keikaku Gakkai-shi*, 23 (4), pp. 265–274.

Kuramoto, N. and Sonoda, Y. (2012). 'Conserving biological diversity', in Duraiappah, A.K., Nakamura, K., Takeuchi, K., Watanabe and M., Nishi, M., *Satoyama-satoumi ecosystems and human well-being: socio-ecological production landscapes of Japan*. Tōkyō: United Nations University Press, pp. 81–89.

Kurita, I. (1983). Akōzantai shin'yō-jurin no seitaigakuteki kenkyū. *Shinrin Ricchi*, 25 (1), 31–37. DOI:10.18922/jjfe.25.1_31.

Lawler, J.J., White, D., Sifnoeos, J.C. and Master, L.L. (2003). Rare species and the use of indicator groups for conservation planning. *Conservation Biology*, 17 (3), pp. 875–882.

Litvaitis, J.A., Sherburne, J.A. and Bissonette, J.A. (1985) Influence of understory characteristics on snowshoe hare habitat use and density. Journal of Wildlife Management, 49(4), pp. 866–873.

Loh, J., Green, R.E., Ricketts, T., Lamoreux, J., Jenkins, M., Kapos, V. and Randers, J. (2005). The living planet index: using species population time series to track trends in biodiversity. *Philosophical Transactions of the Royal Society B: Biological Sciences*, 360, pp. 289–295.

MacArthur, R.H. and MacArthur, J.W. (1961). On bird species diversity. *Ecology*, 42 (3), pp. 594–598.

MacArthur, R.H., and Wilson, E.O. (1967). *The theory of island biogeography*. Princeton: Princeton University Press.

Maruyama, H. (1994). *Kindai nihon kōenshi no kenkyū*. Kyoto: Shibunkaku.

Masaoki, T. (2009). Avifauna of the Nansei Shoto (Southwest Islands), Japan, in relation to distances among islands. *Japanese Journal of Ornithology*, 58, pp. 1–17.

Mathur, P.K., Kumar, H., Lehmkuhl, J.F., Tripathi, A., Sawarkar, V.B. and De, R. (2011). Mammal indicator species for protected areas and managed forests in a landscape conservation area of northern India. *Biodiversity and Conservation*, 20 (1), pp. 1–17.

Matsumoto, S., Yokohari, M., Kato, Y. and Yamamoto, S. (1999). Hatachi keikan hozen notameno chiiki jūmin niyoru keikan no hyōka kōzō no haaku: gunma-ken hokubu no hatasaku-chitai o jirei ni. *Nōson Keikaku Gakkai-shi*, 18 (suppl), pp. 49–54.

Matsumura, T., Takeda, Y. and Nakase, K. (2000). Tanbanomori-kōen niokeru shokusei kanri to shutayōsei no hozen ni kansuru kenkyū. *Kōbe-daigaku Hattatsu-kagakubu Kenkyu-kiyō*, 7 (2), pp. 785–792. *Kōbe University Repository*. Available at: http://www.lib.kobe-u.ac.jp/handle_kernel/81000369 (accessed 12.07.19).

Matsuoka, S., Watari, T., Matsumoto, R. and Shirakawa, H. (2000). The evaluation of sustainability for utilization of forestry resource in Japan: the comparison of property regimes (market, state, and commons) as the sustainable resource management system. *Journal of International Development and Cooperation*, 6 (1), pp. 1–22.

Matsuura, T., Hayashi, M., Sugimura, K., Tanaka, N. and Miyamoto, A. (2013). Sansai kinoko tori ga motarasu seitaikei sābisu no hyōka: Fukushima-ken Tadami-machi o jireini. *Shinrin-keikaku Gakkaishi*, 47, pp. 55–81.

Matsuura, T., Sugimura, K., Miyamoto, A., Tanaka, H. and Tanaka, N. (2014). Spatial characteristics of edible wild fern harvesting in mountainous villages in northeastern Japan using GPS tracks. *Forests*, 5, pp. 269–286.

Mishima, A. (1966). Habu nikansuru kenkyū I, Amami Oshima-san habu no shokusei nitsuite. *Eisei Dōbutsu*, 17, pp. 1–21.

Mitani, N., Morosawa, T., Yamashita, R., Kioka, M., Goto, Y., Hashimoto, T., Kitaura, K., Yamada, F., Abe, S. and Ishikawa, T. (2014). Amami Ōshima no mangūsu taisaku ni dōnyu sareta tansakuken no hyōka. *Yasei Seibutsu to Shakai*, 2 (1), pp. 11–22. Available at: https://www.jstage.jst.go.jp/article/awhswhs/2/1/2_KJ00009778123/_pdf/-char/en.

Mitsui, K. and Ikehara, S. (1979). Yonaha-dake to sono shūhen niokeru nezumi-rui, shokuchū-rui no bunpu. *Okinawa Seibutsu-gakkaishi*, 17, pp. 7–13.

Moriyama, H. (1997). *Mura no shizenn o ikasu*. Tōkyō: Iwanami Shoten.

Motokawa, M., Shimoinaba, S. and Suzuki, S. (2006). Nihonsan honyūrui no saikin no bunrui taikei: Abe (2005) to Wilson and Reeder (2005) no hikaku. *Honyūrui Kagaku*, 42, pp. 181–191. DOI: 10.11238/mammalianscience.46.181.

Murata, G. (2005). Nihon no shokubutsusō to shokuseitai. *Bunrui*, 5 (1), pp. 1–8. DOI: 10.18942/bunrui.KJ00004649627.

Murray D.L., Roth, J.D., Ellsworth, E., Wirsing, A.J. and Steury, T.D. (2002) Estimating low-density snowshoe hare populations using fecal pellet counts. Canadian Journal of Zoology, 80(4), 771–781.

Nagamitsu, T., Nagasaka, K., Yoshimaru, H. and Tsumura, Y. (2014). Kikō-jōken ni kanren shita 50-nensei nihon karamatsu no seichō no sanchikan hen'i. *Shinrin Iden Ikushu*, 3, pp. 111–117. Available at: http://www.fgtb.jp/publish/2014-7.html (accessed 5.07.19).

Nakamura, T., Kuroiwa, A., Nishida, U.C., Matsubara, K., Yamada, F. and Matsuda, Y. (2007). Comparative chromosome painting map between two Ryukyu spiny rat species, *Tokudaia osimensis* and *Tokudaia tokunoshimensis* (Muridae, Rodentia). *Chromosome Research*, 15 (6), pp. 799–806.

Nakayama, T. (1980). 'Shinrin no kōekiteki kinō hyōka', in Kurimura, T. (ed.) *Sanrin no hyōka*. Tōkyō: Nihon Ringyō Gijutsu Kyōkai, pp. 592–629.

Nakazawa, C. (2010). Meiji-jidai Tōhoku-chihō niokeru Nihon ōkami no kujo. *Yaseiseibutsuhogo*, 12 (2), 19–38. DOI: 10.20798/wildlifeconsjp.12.2_19.

Natori, T. (2008). Minami-alupus Kitadake no Kitadake-sō no seiiku ni oyobosu chikyū ondanka no eikyō. *Nihon Seitaigaku Gakkaishi*, 58, pp. 183–189. DOI: .10.18960/seitai.58.3_183.

Neel, M.C. and Cummings, M.P. (2003). Effectiveness of conservation targets in capturing genetic diversity. *Conservation Biology*, 17 (1), pp. 219–229.

Nicholas, V., Brokaw, L. and Lent, R.A. (1999). 'Vertical structure', in Hunter, M.L. (ed.) *Maintaining biodiversity in forest ecosystems*. Cambridge: Cambridge University Press, pp. 373–399.

Nihon Bunruigaku Rengō. (2003). *Nihon-san seibutsu shusū chōsa*. Available at http://ujssb.org/biospnum/search.php (accessed 6.07.19).

Nihon-ginkō. (2020). *Shuyō jikeiretsu dēta-hyō*. Available at: https://www.stat-search.boj.or.jp/ssi/mtshtml/fm08_m_1.html (accessed 19.03.20).

Nihon Yachō-no-kai. (2012). *Fīrudo gaido Nihon no yachō, dai-7han*. Tōkyō: Nihon Yachō-no-kai.

Ninan, K.N. and Inoue, M. (2013). Valuing forest ecosystem services: case study of a forest reserve in Japan. *Ecosystem Services*, 5, pp. 78–87.

Nishikawa, J. (2015). Sado sekai nogyō isan niokeru seibutsu kyōsei nōhō eno torikumi kōka. *Nihon Seitaigaku Gakkaishi*, 65, pp. 269–277. DOI:10.18960/seitai.65.3_269.

Nishikiori, T., Ito, S., Tsuji, H., Yasutaka, T., Hayashi, S. (2015) Rinshō hihuku no chigai ga dojō shinshoku ni tomonau hōshasei seshiumu no idō ni oyobosu eikyō. Nihon Shinrin Gakkaishi, 97 (1), pp. 63–69.

Nonaka, I. (2008). Kokuyūrin wa shisanchōka toiu bakageta taishaku-taishōhyō (kokuyūrin tokushū sono 1). Kokuyūrin-kaikei wa nanio hakatte kitanoka. *Ringyō Keizai*, 61 (3), pp. 9–11.

Nōrin-suisan-shō tōkeijōhō-bu (ed.) (2013). *2000-nen sekai nōringyō sensasu: Ōita-ken tōkeisho (ringyō-hen)*. Tōkyō: Nōrin Tōkei Kyōkai.

Ochiai, S., Ueda S, Hasegawa H. and Kakiuchi H. (2015). Effects of radiocesium inventory on 137Cs concentrations in river waters of Fukushima, Japan, under base-flow conditions. *Journal of Environmental Radioactivity*, 144, pp. 86–95.

Ohnishi, N and Yasukōchi, H. (2010). Kyūshū de saigo ni hokaku sareta tsukinowaguma no kigen. *Honyurui Kagaku*, 50 (2), 177–180. Available at: https://doi.org/10.11238/mammalianscience.50.177 (accessed 18.04.20).

Okada, M. and Asakawa, S. (2001). Heichikei nōson keikan hyōka niokeru jurin no igi to chiikisei. *Landscape Kenkyū*, 64 (5), pp. 741–746.

Okitsu, S. (2003). Forest vegetation of northern Japan and the southern Kurils. *Naganoken-kankyōhozen-kenkyūsho Kenkyū-hōkoku*, 11, pp. 231–261. Available at: https://www.pref.nagano.lg.jp/kanken/johotekyo/kenkyuhokoku/hozen/index.html (accessed 13.12.18).

Oshima, K. (1991). Daiyonki kōki niokeru Nihon Rettō shūhen no kaisui junhendō. *Chigaku Zasshi*, 100 (6), pp. 967–975. DOI: 10.5026/jgeography.100.6_967

Ota, H. (1998). Geographic patterns of endemism and speciation in amphibians and reptiles of the Ryukyu Archipelago, Japan, with special reference to their paleogeographical implications. *Researches on Population Ecology*, 40, pp. 189–204.

Otsuka, K. and Ozeki, M. (2015). Nagano-ken Kiso-chiiki niokeru jōryoku kōyōju shirakashi (buna-ka) no bunpu. *Naganoken-kankyōhozen-kenkyūsho Kenkyū-hōkoku*, 12, pp. 9–13. Available at: https://www.pref.nagano.lg.jp/kanken/johotekyo/kenkyuhokoku/hozen/index.html (accessed 13.12.18).

Ouchi, Y. (1987). Kakudai zōrin seisaku no rekishiteki tenkai katei. *Ringyou Keizai Kenkyū*, 111, pp. 3–11. DOI: 10.20818/jfe.1987.111_3.

Pearman, P.B. (2002). Developing regional conservation priorities using red lists, a hypothetical example from the Swiss lowlands. *Biodiversity and Conservation*, 11 (3), pp. 469–485.

Pearman, P.B. and Weber, D. (2007). Common species determine richness patterns in biodiversity indicator taxa. *Biological Conservation*, 138 (1–2), pp. 109–119.

Penna, I, W. (2002). *The political ecology of the Japanese paper industry*. Ph.D. Dissertation. University of Melbourne.

Prawiradilaga, D.M., Rov, N., Gjershaug, J.O., Hapsoro and Supriatna, A. (2001). Feeding ecology of the Javan Hawk-eagle (*Spizaetus bartelsi*) during the nestling period. *Zoo Indonesia*, 28, pp. 19–25. Available at: e-journal.biologi.lipi.go.id (accessed 4.05.20).

Prendergast, J.R., Quinn, R.M., Lawton, J.H., Eversham, B.C. and Gibbons, D.W. (1993). Rare species, the coincidence of diversity hotspots and conservation strategies. *Nature*, 365, pp. 335–337.

Ridder, B. (2008). Questioning the ecosystem services argument for biodiversity conservation. *Biodiversity and Conservation*, 17, pp. 781–790.

Rin'ya-chō (1957–2018) *Shinrin ringyō tōkei yōran (issued annually)*. Tōkyō: Rin'ya-kōsaikai.

Saito, H. (2006). *Jikyūteki oyobi rekuriēshonteki na sansai kinoko tori nikansuru kenkyū: saishu shigen no seikaku to seitaiteki haikei (hakase-ronbun)*. Kyōto: Kyōto Daigaku.

Saito, O. and Shibata, H. (2012). 'Satoyama and satoumi, and ecosystem services: a conceptual framework', in Duraiappah, A.K., Nakamura, K., Takeuchi, K., Watanabe, M., and Nishi, M., *Satoyama-satoumi ecosystems and human well-being: socio-ecological production landscapes of Japan*. Tōkyō: United Nations University Press, pp. 17–59.

Sakamoto, T. (1994). Seihukan zaiseikankei no kaikaku to bunkenka: chihō zeisei no kaikaku to chihō shōhizei dōnyu mondai. *Okayama-daigaku Keizaigakkai Zasshi*, 26 (2), pp. 1–22. Available at: http://ousar.lib.okayama-u.ac.jp/files/public/4/41709/2016052804095657750/oer_026_2_001_022.pdf (accessed 15.07.19).

Sakata, T., Iwamoto, T. and Baba, M. (2013) Kyūshu no kamoshika: zetsumetsu no kiki o kaihi suruniwa. *Dai-29-kai Nihon reichōrui-gakkai Nihon honyūrui-gakkai 2013-nendo gōdōtaikai*, P-140. Available at: https://www.jstage.jst.go.jp/article/primate/29/0/29_221_1/_pdf/-char/ja (Accessed: 18.04.20).

Sanada, Y., Katada, G. and Kaneyasu, N. (2018). Sanchi no kumo ya kiri ga motarashita hōshanō osen o kaimei: kōkūki monitaringu to sūchi simyureishon niyoru kaiseki. *Aisotōpu Nyūsu*, 759, pp. 18–21.

Sanbe, H. (2018). "Shakaiteki kikan" toshiteno kigyō no CSR. *Keieigaku Ronshu*, 88, pp. 65–71.

Sasaoka, T., Shoji, K., Torii, T. and Hashimoto, Z. (1995). Todōhuken-ritsu sizehkōen niokeru hozen no jittai to bunseki. *Landosukēpu Kenkyū*, 58 (5), pp. 237–240. DOI: 10.5632/jila.58.5_237.

Sato, H., Goto, H., Kotaka, N., Sueyoshi, M., Nomiya, H., Tanouchi, H., Sugimura, K., Neda, H., Abe, S., Hasegawa, M., Hattori, T., Saitō, K. and Yamada, F. (2011). Okinawa Yambaru chiiki no mori no riyō to seibutsutayōsei. Shinrin Sōgō- kenkyūsho Heisei 22 nendo Kenkyūseika Senshū, pp. 18–19. Available at: https://www.ffpri.affrc.go.jp/pubs/seikasenshu/2010/documents/p18-19.pdf (accessed 17.04.20).

Schmidt, I., Zerbe, S., Betzin, J. and Weckesser, M. (2006). An approach to the identification of indicators for forest biodiversity. *Restoration Ecology*, 14 (1), pp. 123–136.

Shackleton, S., Shackleton, C. and Shanley, P. (eds.) (2011). *Non-timber forest products in the global context*. Berlin Heidelberg: Springer.

Shigematsu, Y., Sugimura, K. and Sakai, M. (2018). Fukushima-ken niokeru shizen taikengata rekuriēshon no henka: higashi-nihon daishinsai no eikyō o saguru. *Nagasaki-daigaku Sōgō Kankyō Kenkyū*, 21, pp. 43–54. Nagasaki University's Academic Output SITE: http://naosite.lb.nagasaki-u.ac.jp/dspace/handle/10069/38720?locale=ja.

Shimizu, Y., Yahara, T. and Sugimura, K. (1988). Amami Ōshima niokeru anettai jōryoku kōyōjurin no bassai-go no sen'i katei. *Komazawa-chiri*, 24, pp. 31–56.

Shinjo, Y. and Nishimura, M. (1984). Minami takeichirō-shi no hokaku keiken kara mita habu no seitai. *Okinawa Seibutsu-gakkaishi*, 22, pp. 115–116.

Sōmushō Tōkeikyoku. (2017). *Heisei 27-nen kokusei chōsa*. Available at: https://www.stat.go.jp/data/kokusei/2015/kekka.html (accessed 18.07.19).

Sprague, D.S., Goto, T. and Moriyama, H. (2000). Jinsoku sokuzu no GIS kaiseki niyoru Meiji shoki no nōson tochi riyō no bunseki. *Landscape Kenkyū*, 63 (5), pp. 771–774.

Stattersfield, A.J., Crosby, M.J., Long, A.J. and Wege, D.C. (1998). *Endemic bird areas of the world: priorities for biodiversity conservation*. Cambridge: Cambridge University Press.

Sugimura, K. (1987). *Forestry and wildlife conservation on Amami Oshima, Japan: an integrated study of wildlife and human society*. Ann Arbor: University Microfilm International.

Sugimura, K. (1988). The role of government subsidies in the population declines of some unique wildlife species on Amami Oshima, Japan. *Environmental Conservation*, 15 (1), pp. 49–57.

Sugimura, K. (1990). Upsurge of environmental concern and the current attitude towards conservation in Japan. *Environmental Conservation*, 17 (3), pp. 270–272.

Sugimura, K. (1993). Amami-no-kurousagi no seitai, bunpu to seisokusū no henka nitsuite. *Chirimosu*, 4, pp. 5–11.

Sugimura, K. (2000). 'The Amami rabbit *Pentalagus* furnessi', in Chapman, J.A. and Flux, J.E.C. (eds.) *Rabbits, hares and pikas: status survey and conservation action plan.* Gland: Switzerland, pp. 141–143.

Sugimura, K. (2002). Amami Ōshima niokeru chōrui to Amami-no-kurousagi no seisokusū no hendō keikō to kankyō henka nitsuite. *Kankyō Johō Kagaku Ronbushū*, 16, pp. 121–126.

Sugimura, K. (2012). Kanbenna teiryōteki shihyō o mochiita shinrin no kachi oyobi riyō jittai no chiikikan hikaku. *Kankyo Johō Kagaku Gakujutsu Ronbunshū*, 26, pp. 313–318.

Sugimura, K. (2016). Amami Oshima niokeru ringyō, tsūrizumu oyobi kishōshu hozen no kyōzon notameno shirin kanri hōshin nitsuite. *Kankyo Johō Kagaku Gakujutsu Ronbunshū*, 30, pp. 145–150.

Sugimura, K. and Howard, T. (2008). Incorporating social factors to improve the Japanese forest zoning process. *Forest Policy and Economics*, 10 (3), pp. 161–173.

Sugimura, K. and Yamada, F. (2004). Estimating population size of the Amami rabbit *Pentalagus furnessi* based on fecal pellet counts on Amami Island, Japan. *Acta Zoologica Sinica*, 50 (4), pp. 519–526.

Sugimura, K., Ishida, K., Abe, S., Nagai, Y., Watari, Y., Tatara, M., Takashi, M., Hashimoto, T. and Yamada, F. (2013). Monitoring the effects of forest clear-cutting and mongoose *Herpestes auropunctatus* invasion on wildlife diversity on Amami Island, Japan. *Oryx*, 48 (2), pp. 241–249.

Sugimura, K., Matsuura, T. and Miyamoto, A. (2007). Recreational use of roads in the satochi landscapes in southern Ibaraki prefecture. *Journal of Environmental Information Science*, 35 (5), pp. 119–126.

Sugimura K, Sahab, A., Yata, M., Kridalaksana, A., Zanuansyah, A., Ichwani, S.N., Nurika, S. and Howard, T.E. (2015). Local people's use of non-timber forest products in the Gunung Halimun Salak National Park, West Java. *Nagasaki-daigaku Sōgō Kankyō Kenkyū*, 18 (1), pp. 16–27. Available at: https://irdb.nii.ac.jp/01160/0001730076 (accessed 29.04.20).

Sugimura, K., Sato, S., Yamada, F., Abe, S., Hirakawa, H. and Handa, Y. (2000). Distribution and abundance of the Amami rabbit *Pentalagus furnessi* in the Amami and Tokuno Islands, Japan. *Oryx*, 34 (3), pp. 198–206.

Suzuki, M. and Oumi, S. (2017). Amami Ōshima no kajuen no dōitsu hanshoku-ana niokeru amaminokurousagi no hanshoku to yōikukōdo no 2-jirei. *Honyūrui Kagaku*, 57 (2), pp. 257–266.

Tadami-chōshi Hensan-iinkai (2001). *Tadami-machi no shizen: kikō, chishitsu, dōbuts, shokubutsu*. Tadami: Fukushima-ken Tadamai-machi.

Tadami-machi Kyōiku-iinkai (2003) *Fukushima-ken Tadami-chiku shinrin-shokusei to seibutsu-tayōsei gakujutsu-chōsa hōkokusho, tokuni Fagus crenata to Pterocarya rhoifolia no identeki kōzō nitsuite*. Tadami: Tadami-machi.

Tadami-machi-yakuba Sōmuka. (2015). *2014 Tadami chōsei yōran*. Hūkōsha. Available at: https://www.town.tadami.lg.jp/Files/2015/04/09/Handbook.pdf (accessed 15.12.19).

Takacs, D. (1996). *The idea of biodiversity: philosophies of paradise*. Baltimore and London: Johns Hopkins University Press.

Takahashi, M. and Matsuura, T. (2011). Shijō ni okeru sansai torihiki no jittai nitsuite: sapporo-shi chuō oroshiuri shijō o reini. *Nihon Shinringakkai Hokkaidō-shibu Ronbunshū*, 59, pp. 113–116. DOI: 10.24494/jfsha.59.0_113.

Take, M. and Saito, K. (2011). Bunken niyoru ekotsūrizumu niokeru gaido no yakuwari to kankyōhozen tono kankei haaku. *Landoscēpu Kenkyū*, 74 (5), pp. 531–536 DOI: 10.5632/jila.74.531

Takeuchi, K. (2003). 'Satoyama landscapes as managed nature', in Takeuchi, K., Brown, R.D., Washitani, I., Tsunekawa, A. and Yokohari, M. (eds.) *Satoyama: the traditional rural landscape of Japan*. Tōkyō: Springer, pp. 9–16.

Takeuchi, K., Brown R.D., Washitani, I., Tsunekawa, A. and Yokohari, M. (eds.) (2003). *Satoyama: the traditional rural landscape of Japan*. Tōkyō: Springer.

Taki, H., Okabe, K., Makino, S., Yamaura, Y. and Sueyoshi, M. (2009). Contribution of small insects to pollination of common buckwheat, a distylous crop. *Annals of Applied Biology*, 155, pp. 121–129.

Tamaki, Y., Sajiki, T. and Kubota, H. (2016). Keiryū tsuriba no zōning kanri ni taisuru tsurikata betsu yūgyosha no shikō no sai. *Rikusui-gaku Zasshi*, 77, pp. 145–153 DOI: 10.3739/rikusui.77.145.

Tanaka, T. (2012). 'Yowai chiikisei' toshiteno Nihon no kokuritsu kōenseido: gyōsei bumon niokeru shigen to kengen no kokusai hikaku. *Shinsedai Hōseisakugaku Kenkyū*, 17, pp. 369–402. Available at: http://hdl.handle.net/2115/49999. (accessed 31.07.19).

Tanokura, N., Yokohari, M., Yamamoto, S. and Kato, Y. (1999). Jimoto jūmin niyoru suiden keikan no ninchi kōzō. *Landscape Kenkyū*, 62 (5), pp. 727–732.

Terauchi, M., Nakamura, K., Matsuoka, S. and Miyashita, N. (1985). Nōkōchi niokeru kijibato kotaisū no kisetsuhenka to soreni kan'yosuru esajōken. *Tori*, 34 (1). 7–16. DOI:10.3838/jjo1915.34.7.

Tilman, D. (2001). 'Functional diversity', in Levin S. (ed.) *Encyclopedia of biodiversity*. San Diego: Academic Press, pp. 109–121.

Tojo, H. (2007). Nihon-san shinrin izonsei chōrui shusū no suitei. *Shinrin Sōgō-kenkyūsho Kenkyū Hōkoku*, 6 (1), pp. 9–36. Available at: https://www.ffpri.affrc.go.jp/labs/kanko/402-2.pdf (accessed 5.05.20).

Tokuda, M. (1969). *Seibutsu chirigaku*. Tōkyō: Tsukiji Shokan.

Tsukuba-shi. (2017). *Dai-niji Tsukuba-shi kankō kihon keikaku*. Available at: https://www.city.tsukuba.lg.jp/kankobunka/kankojoho/1001464.html (accessed 28.12.18).

Tsunekawa, A. (2003). 'Transition of satoyama landscapes in Japan', in Takeuchi, K., Brown, R.D., Washitani, I., Tsunekawa, A. and Yokohari, M. (eds.) *Satoyama: the traditional rural landscape of Japan*. Tōkyō: Springer, pp. 41–51.

Ueda, A. and Onodera, H. (2004). Satochi-satoyama to seibutsu tayōsei. *Nōson Kenkyū*, 98, pp. 25–33. Available at: http://nodaiweb.university.jp/noukei/pdf/NSO98_03.pdf (accessed 1.08.18).

Ueda, T., Fujii, N. and Takeuchi, T. (1993). Baburu Keizai hōkaigo no shōhisha kōdō. *Gakushūin-daigaku Keizai-ronshū*, 30 (1), pp. 1–51.

Ueno, Y. (2009) Yairochō. *Bādo Risāchi Nyūsu*, 6 (9), p. 4.

Ueyama, S. (1969). *Shōyōjurin bunka: Nihon bunka no shinsō*. Tōkyō: Chūō-Kōron-sha.

United Nations (2019). *World Population Prospects 2019*. Available at https://population.un.org/wpp/ (accessed 17.07.19).

Usuki, M. (2004). Present State and perspectives on national parks and protected areas in South-East Asian region. *Tropics*, 13 (3), pp. 221–232.

Velazquez, A. and Heil, G.W. (1996). Habitat suitability study for the conservation of the volcano rabbit (*Romerolagus diazi*). *Journal of Applied Ecology*, 33, pp. 543–554.

Walker, B.L. (2009). *The lost wolves of Japan*. Seattle: University of Washington Press.

Wallace, C. (2007). Classification of ecosystem services; problems and solutions. *Biological Conservation*, 139 (3–4), pp. 235–246.

Watari, Y. (2008). *Gairaishu Jawa mangūsu ga Amami Ōshima no zairai seibutsu gunshū ni oyobosu eikyō to sono kikō no kaimei*. Ph.D. Dissertation, Tōkyō University.

Watari, Y., Nishijima, S., Fukasawa, M., Yamada, F., Abe, S. and Miyashita, T. (2013). Evaluating the "recovery level" of endangered species without prior information before alien invasion. *Ecology and Evolution*, 3 (14), pp. 4711–4721.

Watari, Y., Takatsuki, S. and Miyashita, T. (2008). Effects of exotic mongoose (*Herpestes javanicus*) on the native fauna of Amami-Oshima Island, southern Japan, estimated by distribution patterns along the historical gradient of mongoose invasion. *Biological Invasions*, 10 (1), pp. 7–17.

Williamson, M. (1989). Natural extinction on islands. *Philosophical Transactions of the Royal Society of London B: Biological Sciences*, 325(1228):457–66. DOI: 10.1098/rstb.1989.0099.

Xu, X., Hirata, E., Tokashiki, Y. and Shinohara, T. (2001). Structure and species diversity of subtropical evergreen broad-leaved forest in northern Okinawa Island, Japan. *Journal of Forest Research*, 6 (3), pp. 203–210.

Yamada, A. (2002). Nihon-san kinkon-sei kinoko-rui no shoku-shigen toshiteno riyōsei. *Shinshū-daigaku Nōgakubu Kiyō*, 38 (1–2), pp. 1–17.

Yamada, F. and Sugimura, K. (2004). Negative impact of an invasive small Indian mongoose *Herpestes javanicus* on native wildlife species and evaluation of a control project in Amami-Ohshima and Okinawa Islands, Japan. *Global Environmental Research*, 8 (2), pp. 117–124.

Yamada, F., Takenoshita, Y., Nakatani, J., Kawamura, S., Ōi, T., Ōtsuki, K., Konno, F., Hayama, S. and Horino, S. (2013). Fukushima genpatsu jikogo no hōshanō eikyō o ukeru yasei honyūrui no monitaringu to kanri mondai ni taisuru teigen. *Honyūrui Kagaku*, 53 (2), pp. 373–386.

Yamaguchi, Y., and Yoshida, S. (2006). Suitō jikamaki saibai niokeru kijibato no higaitaisaku toshiteno daitaiesa no kōka. *Nihon Chō-gakkaishi*, 55 (1), 1–6. DOI:10.2326/jjo.55.1

Yamashita, S., Okabe, K. and Sato, T. (2013). Shinrin seitaikei niokeru seibutsu-tayōsei to tanso junkan. *Shinrin-sōgō kenkyūsho Kenkyū-hōkoku*, 12 (1), pp. 1–21. Available at: https://www.ffpri.affrc.go.jp/pubs/bulletin/426/index.html (accessed 11.07.19).

Yamazaki, H., Aoki, K., Hattori, T. and Takeda, Y. (2000). Satoyama no shokusei kanri niyoru shutayōsei no zōka. *Landosukēpu Kenkyū*, 63 (5), pp. 481–484. DOI: 10.5632/jila.63.481.

Yokoyama, R. (2006) Fukushima-ken Aizu-chiiki ni Nihon saidai no sihinrin hogoku ga tsukurareru kotoni narimashita. Shizenhogo, 491, p. 37.

Yoshida, N. and Kanda, J. (2012). Tracking the Fukushima radionuclides. Science, 336, pp. 1115–1116.

Yoshihara, T., Matsumura, H., Hashida, S. and Nagaoka, T. (2013) Radiocesium contaminations of 20 wood species and the corresponding gamma-raydose rates around the canopies at 5 months after the Fukushima nuclear power plant accident. *Journal of Environmental Radioactivity*, 114, pp. 60–68.

Yoshimura, M. (2010) Comparison of EPT (Ephemeroptera, Plecoptera, Trichoptera) adult assemblage between old-growth natural forest and planted coniferous forest basins in Japanese temperate region. Research Report of the Research and Education Center for Inlandwater Environment, Shinshu University, 6, pp. 83–93.

Yasuda, M. (2007). Zetsumetsu no osore no aru Kyūshū no nihonrisu, momonga oyobi musasabi: kako no seisoku kiroku to genjō oyobi kadai. *Honyūrui Kagaku*, 47 (2), pp. 195–206. Available at: https://www.jstage.jst.go.jp/article/mammalianscience/47/2/47_2_195/_pdf/-char/ja (accessed 19.04.20).

Yasuda, M., Yayota, T. and Kurihara, T. (2013). Iya-sankei niokeru jidō satsukei kamera de akirakaninatta chūōgata honyūrui no bunpu. *Dai-29-kai Nihon reichōrui-gakkai Nihon honyūrui-gakkai 2013-nendo gōdōtaikai*, F1-5. Available at: https://doi.org/10.14907/primate.29.0_126_2 (accessed 18.04.20).

Index

For Product Safety Concerns and Information please contact our EU representative GPSR@taylorandfrancis.com
Taylor & Francis Verlag GmbH, Kaufingerstraße 24, 80331 München, Germany

www.ingramcontent.com/pod-product-compliance
Ingram Content Group UK Ltd.
Pitfield, Milton Keynes, MK11 3LW, UK
UKHW020005200926
475702UK00015B/2495
9780367634964